RACE, POLITICS, AND RECONSTRUCTION

The Black Soldier in War and Society

New Narratives and Critical Perspectives

Edited by LE'TRICE D. DONALDSON AND GEORGE WHITE JR.

RACE, POLITICS, AND RECONSTRUCTION

THE FIRST BLACK CADETS
AT OLD WEST POINT

Edited by Rory McGovern and
Ronald G. Machoian

University of Virginia Press
CHARLOTTESVILLE AND LONDON

University of Virginia Press
© 2024 by the Rector and Visitors of the University of Virginia
All rights reserved
Printed in the United States of America on acid-free paper

First published 2024

1 3 5 7 9 8 6 4 2

LIBRARY OF CONGRESS CATALOGING-IN-PUBLICATION DATA
Names: McGovern, Rory, editor. | Machoian, Ronald Glenn, editor.
Title: Race, politics, and Reconstruction : the first Black cadets at old West Point / edited by Rory McGovern and Ronald G. Machoian.
Description: Charlottesville : University of Virginia Press, 2024. | Series: The Black soldier in war and society : new narratives and critical perspectives | Includes bibliographical references and index.
Identifiers: LCCN 2024019024 (print) | LCCN 2024019025 (ebook) | ISBN 9780813951904 (hardback) | ISBN 9780813951911 (paperback) | ISBN 9780813951928 (ebook)
Subjects: LCSH: United States Military Academy—Admission—History—19th century. | African American military cadets—History—19th century. | Military cadets—United States—History—19th century. | United States—Armed Forces—African Americans—History—19th century. | Reconstruction (U.S. history, 1865–1877) | United States—Race relations—History—19th century. | United States Military Academy—Public opinion—History—19th century. | Public opinion—United States. | BISAC: HISTORY / United States / State & Local / South (AL, AR, FL, GA, KY, LA, MS, NC, SC, TN, VA, WV)
Classification: LCC U410.Q1 R34 2024 (print) | LCC U410.Q1 (ebook) | DDC 355.0071/173—dc23/eng/20240509
LC record available at https://lccn.loc.gov/2024019024
LC ebook record available at https://lccn.loc.gov/2024019025

Cover art: Henry O. Flipper as a cadet. (United States Military Academy Library Special Collections and Archives)
Cover design: David Fassett

CONTENTS

Illustrations follow page 110

RACE, POLITICS, AND RECONSTRUCTION

INTRODUCTION

B ut for one glaring exception, the graduation ceremony at West Point on June 14, 1877, was much like others in years past. Spectators gathered in a hollow square in the shade of some maple trees thriving in the summer weather near the United States Military Academy's lone academic building. At 10:30 a.m. the band struck up a lively martial tune and the graduating class of 1877 marched the short distance from their barracks to the assembled crowd, taking their seats within the center of the square. They listened in rapt attention as Maj. Gen. Winfield Scott Hancock delivered a graduation speech that welcomed them to the United States Army officer corps, enjoined them to be lifelong students of their profession and of any other fields of knowledge that called to them, and challenged them to seek active service in the toughest frontier assignments. The secretary of war and the superintendent of West Point then delivered their own perfunctory and congratulatory remarks before finally beginning to award diplomas. The first went to William M. Black, a future chief engineer of the US Army who graduated at the top of his class. Commanding General William T. Sherman rose from his seat and warmly congratulated Black's understandably proud father.[1]

When it came time for Henry O. Flipper to accept his diploma, the ceremony was no longer a mere repetition of past graduations with new faces and names taking their places in the army. Flipper had persevered through four long years of soul-crushing isolation and ill treatment to become West Point's first Black graduate. His presence at the academy, like that of several Black cadets who preceded him, was contested and controversial from the day he arrived. Friends and foes alike watched, discussed, and wrote about his progress, trials, and triumphs. Everyone in attendance knew at least some of what he had endured in those four years, Sherman not least among them. On hearing Flipper's name, according to a correspondent for the *New York Herald,* Sherman "clapped his hands approvingly and his example was at once followed by all the visitors and officers present, until there was an almost universal round of applause." No other graduating cadet enjoyed the same spontaneous recognition

that day, not even William Black as valedictorian. Flipper quickly bowed his
head in silent acknowledgement, basking for the briefest of moments in the
applause.[2] He must have felt a strange combination of pride, hope, and regret:
pride in his accomplishments, hope for his future and for the futures of other
young Black men who wanted to follow in his footsteps, and regret that he had
to wait until graduation day for the West Point community to acknowledge
him in any meaningfully positive way.

Flipper richly deserved Sherman's ovation. The administration of President Ulysses Grant professed a commitment to integrating West Point and
enjoyed strong support from Radical Republicans to do so. While Grant's
policy and its congressional advocates did much to get Black cadets to West
Point, neither they, the army, nor West Point officials did much, if anything,
to ensure that Black cadets would receive fair treatment once there and have
a reasonable chance to graduate. Graduation rates reflected that fact. Flipper
was the sixth Black nominee and the fourth Black cadet admitted to West
Point since Grant was sworn into office in 1869. West Point's Academic Board
had either refused to admit or dismissed all but Flipper after finding them deficient in entrance, midyear, or year-end examinations. During Flipper's four
years at West Point, there were an additional five Black nominees, but the
Academic Board admitted only two, neither of whom graduated. As Reconstruction ended with the swearing in of a new president and a new congress
in the same year that Flipper graduated, political will to sustain integration
at West Point began to fade. West Point admitted six Black cadets in the decade following Flipper's graduation. Only two—John Hanks Alexander and
Charles Young—graduated, and there would not be another until 1936.[3] The
first and ultimately failed attempt to integrate West Point was a function of
the complex intersection of the political purposes, racial prejudices, and social
pressures that defined Reconstruction-era America.

This volume analyzes the integration of West Point as both a human experience and a process framed and driven by its unique Reconstruction-era
social and political contexts. This is an approach that differs from most other
historical examinations of integration at West Point during and after Reconstruction. It is an approach that is necessary to better understand not only how
those involved experienced and interpreted integration at West Point but also
how and why the process of integration took the form and path that it did.

Past analyses of this first attempt to integrate West Point usually take one of
two forms, the first of which is biographical. In these studies, an individual's
experience of or relationship to integration at West Point during and after Reconstruction is the focus, not the process of integration itself. These studies are

rich in human interest details and do much to illuminate how specific Black cadets and leaders at West Point experienced and interacted with integration at the academy.[4] Biographical studies of this genre make valuable contributions to our knowledge of integration at West Point but focus much more on the human experience of integration than on its broader social and political processes. Admittedly, the very best among these use biographic narratives as a vehicle to make broader connections to wider social and political currents driving integration at West Point.[5] At the same time, biographical work that covers integration at West Point tends to focus almost exclusively on the cases of Flipper, Johnson C. Whittaker, and Charles Young. All three are exceptional cases for different reasons and therefore are difficult to consider as entirely representative of the Black experience of Reconstruction-era West Point even though elements of each of their stories align well with the average Black cadets' experience.

The other dominant form is that of institutional history. In these studies, integration at West Point appears as a chapter at best or, more likely, just a few paragraphs within the broader history of the academy or the US Army. In this context, Reconstruction-era integration at West Point is presented as an interesting but unfortunately doomed effort at a time in its history generally defined by indiscipline and institutional stagnation.[6] Institutional histories often attach the tortured path of integration at West Point during and after Reconstruction to flagging discipline among cadets—and even in some cases to overassertiveness on the part of certain Black cadets, most often James W. Smith. They generally characterize West Point's faculty and leaders as dutiful officers who moved beyond the prejudices of their day to treat Black cadets justly and fairly as executive policy required. Such interpretations badly need correction. They rely heavily on two influential histories of West Point written during the 1960s, in an environment framed by prevalent white, middle-class views of the modern civil rights movement. These accounts also accepted outright Flipper's published commentary about fair treatment from West Point faculty—at a time when asserting otherwise in a public forum would have jeopardized his career.[7]

Understanding Reconstruction-era integration at West Point as both an experience and a process depends above all on due consideration of broader context. Integration was neither entirely about West Point nor entirely about those who experienced integration at West Point. Politics, legacies of the Civil War, perceptions of race, the experiences of those involved, and the observations and responses of those who had an interest in the outcome all framed and affected the course of events. The first attempts to integrate Black cadets at West Point are part of a broader national narrative that defies telling from any singular historical perspective. The essays here carry a central message: the historian's essential truth that *context matters*.

We organized this volume to analyze Reconstruction-era integration at West Point as a process inextricably bound by the broader context of a national contest over Black rights, roles, and participation in American society after the Civil War and emancipation. In this way, we present a more complete accounting of West Point's first experience with integration—an experience lived not only by Cadet James Webster Smith and those Black cadets who followed him but by the country as well. This volume approaches Black cadets' arrival at West Point from several distinct vantage points to enrich a story that defies any singular perspective.

The volume opens with a contextual framework. Adam H. Domby places West Point within the period's broader political landscape, analyzing it as a federal institution that was both subject and participant in the national discourse of race and Reconstruction. In the second essay, Cameron D. McCoy frames integration at West Point as an extension of the legacy of Black service in the Civil War, which strengthened Black Americans' resolve that emancipation would be a beginning rather than an end and a means to realize freedom in its totality—replete with all of its opportunities for advancement.

The volume then turns to the experiences and perceptions of those who lived through integration at West Point. The third essay, cowritten by Makonen Campbell and Louisa Koebrich, presents a collective history of the several Black men nominated to West Point during Reconstruction, weaving their personal stories into a valuable whole that transcends that of each alone. Campbell and Koebrich assert that the earliest Black cadets adroitly built on the successes and failures of one another, very much aware that they would have to find pathways to succeed despite reactionary obstacles placed in their way. Ronald G. Machoian's essay then provides an intimate analysis of one white cadet's responses and reactions to James Webster Smith's cadet career. This essay reveals first-person insight into the racialized resentment for Smith's arrival among the white corps of cadets, presenting cadet culture itself as a trammel that isolated and demeaned Black cadets, helping bring on their ultimate failure. The fifth essay builds on that foundation, examining responses and resistance to integration among white cadets and faculty. Rory McGovern traces a culture of resistance within the corps of cadets that was aided and abetted by bias and resistance among the faculty.

The volume then considers reactions and responses to integration from three constituencies beyond West Point. In his essay, Jonathan D. Bratten sheds considerable and sometimes surprising light on responses from the broader US Army by mining perspectives that officers shared in contemporary professional journals as well as private and public correspondence. In a similar vein, Amanda M. Nagel examines white public response to the introduction of Black cadets at West Point. Through Reconstruction-era newspaper commentary, Nagel finds and presents a form of collective resistance to Black claims on the institutions and opportunities so jealously guarded by a white society ill-prepared to accept true Black social equality. Le'Trice Donaldson's contribution, presented as this book's final essay, examines Black Americans' support for integration at West Point as a powerful and resolute claim to those opportunities that whites would deny them despite emancipation's promise. Donaldson's essay is a fitting bookend to the volume, shedding light on Black perseverance at the outset of what would become a long and arduous fight for equality, in every meaning and implication of the term.

Collectively, these essays are the story of a national trauma, bringing the complex intersection of race, politics, and Reconstruction into focus on West Point's stage as an expression of a period that promised and ultimately reneged on Black equality. Without this contextual framework, the individual stories of Reconstruction-era West Point's Black cadets are denied their collective significance. In a time defined by social progress and retreat, the adage of "one step forward and two steps back" applies with discouraging accuracy to the first attempts to integrate the US Military Academy and thus also the army officer corps.

Two central themes appear throughout the volume. First, Black Americans viewed and understood military service, both during and after the Civil War, as a path toward true social equality. The equality they desired, and even demanded with their own blood sacrifice, transcended emancipation alone. Through military service, Black Americans reclaimed their humanity by assuming agency over their own lives and the destinies of their families in ways previously denied not only to enslaved people but also, in large degree, to many of those who lived in freedom in the antebellum North. After the war and during Reconstruction, West Point and an officer's commission represented the promise of social leadership—an earned place that implied some measure of social equality. This is an important realization that helps refine our modern understanding of why Black matriculation and success at West Point held forth something more significant than just the advent of institutional integration itself. If higher education was a step toward recognition of one's social achievement, then graduation from West Point as an army officer was a shining trophy on that same path.[8]

The second theme highlights the truth that most white Americans of the period—even the northern public—considered Black freedom and Black equality to be separate and distinct issues. Furthermore, they did not believe that freedom necessarily led to social equality. By the end of the war, most northerners, and even a fair proportion of southerners, were willing to concede at least the principle of civic or political equality under the law for Black Americans. This fueled public acceptance of emancipation and the Thirteenth Amendment, and something between acquiescence to and acceptance of the Fourteenth Amendment; but it also fueled public controversies over Black suffrage, education, land ownership, officeholding, and more. Because of racism's pervasive reach, social equality—generally defined as equality of access to public spaces, institutions, and levers of power—was much more controversial and bitterly contested.[9]

This was true even among veteran officers. There were abolitionists among them who were zealots for antislavery and equality. And there were abolitionists among them who believed slavery was a moral stain on the United States but who also believed Black Americans to be part of a socially and intellectually inferior race. There were far more still who at some point after the Emancipation Proclamation had to come to terms with the fact that they were risking their lives in the cause of emancipation. Accordingly, many white officers and likeminded civilians jealously guarded as their own prerogative the very access to an officer's commission that Black Americans came to view as both a symbol and guarantor of true citizenship. This is why so many officers and cadets, even those who might have cheered the Civil War's outcome and the end of slavery, were not prepared to place Black Americans on a truly equal footing, especially not in their own midst at West Point.[10]

Understanding how these themes affected the US Army and the US Military Academy requires some grounding in the state of the army and West Point at the time. During Reconstruction, the army was oversaturated with missions that competed for scant human and material resources. Its transition from war to peace was incomplete at best, a condition that accurately reflected the environment it operated within even as Congress calibrated its troop strength for a peacetime footing. The Reconstruction-era army occupied and, for a time, governed the South. There, it helped staff and supply the Freedmen's Bureau and enforced—where possible—the Fourteenth and Fifteenth Amendments, the Enforcement Acts, and the Ku Klux Klan Act. At the same time, missions on the frontier multiplied as the line of settlement pressed westward. Beyond the many traditional requirements of a multifunctional frontier constabulary, the Reconstruction-era army deployed a large force to the Mexican border amid a standoff with the French puppet regime in Mexico; waged wars of expansion against the Comanche, Apache, Sioux, Cheyenne, Ute, Paiute, and Modoc nations, among others; and performed various security and logistical roles related to the construction of the transcontinental railroad. On top of this, the army maintained its traditional coastal defense mission, committed engineers to river and harbor improvement work on inland and coastal waterways, continued to explore and map the American West, and oversaw the development of national cemeteries and the subsequent grisly process of disinterring over 300,000 Union war dead from old battlefields for reinterment on consecrated ground.[11]

The army's expanding requirements seemed to belie its shrinking size. At the end of the Civil War in May 1865, the army's muster rolls included 1,000,516 officers and enlisted soldiers. By 1867, the army's authorized end strength had dwindled to 57,194, but it could only recruit enough to fill 56,815 of those billets. And the army had to disburse these soldiers in over 250 military posts to meet its requirements. In many cases, the total military power at any one post amounted to dozens of soldiers rather than hundreds. The Reconstruction-era army would never be larger than it was in 1867. By the latter half of the 1870s, the army fluctuated between only 24,140 and 28,565 officers and enlisted soldiers.[12]

The disparity between demands and resources forced army leaders to prioritize their time and effort. Ultimately, the frontier mission grew until it consumed most of the army's attention and resources, leaving other operations and initiatives scaled back or withering on the vine. The army's presence in the South was an impressive 18,000 in 1869 and remained strong enough to actively fight the Ku Klux Klan into the early 1870s, but forces eventually dwindled to a nadir of 3,000 soldiers in 1876. A plan of organization that allowed for six segregated regular army regiments manned by Black soldiers was trimmed to four—two each of infantry and cavalry. Ideas for professional reforms in training, doctrine, and military education for the most part had to keep percolating throughout Reconstruction before the institution regained sufficient capacity to act upon them in the 1880s and beyond. West Point, too, suffered as the army focused its attention and resources on ever-expanding missions across the West.[13]

It was within this setting that two young Black men arrived at West Point in late May 1870, appointed to the United States Corps of Cadets by Republican congressmen who wished to establish Black Americans' right to attend the prestigious institution. Michael Howard failed the entrance examination and was sent home, but when James Webster Smith passed the exam and remained to matriculate with his class, his presence took on a symbolic significance far beyond the success or failure of one cadet. West Point became yet another stage in the battle between those who clung to the antebellum past replete with racial slavery and those who looked to a reconstructed future characterized by varying degrees of racial equality. The outcome should have been a foregone conclusion with President Grant in the White House, but even with his support, there were limits to federal authority. Executive policies and even federal law could express intent, but giving them shape and enforcing them was another matter. This was the first time that a Black man donned cadet gray and attempted to join the corps of cadets, but it was not West Point's first exposure to socially charged issues that pulled it into debate beyond its stone edifice.[14]

American colleges and universities had been a touchstone for broader political contention since the early nineteenth century, drawing criticism from Jacksonian Democrats as a supposed bulwark for elitist spirit in a meritocratic society. West Point, as a federal academy for young men who would lead the small regular army, had attracted its fair share of such criticism. Later, the fact that so many graduates had resigned from the army to accept wartime Confederate commissions was not missed by many. When northern armies suffered early battlefield defeats at the hands of southerners led by West Pointers in rebellion, it only fanned arguments that the academy may have taught cadets the military arts but failed to inculcate undivided loyalty to the nation. Later in the war, victories brought by graduates at the head of Union armies dampened criticism but never completely mollified critics. Historian W. Scott Dillard aptly characterized the Civil War's legacy at West Point as one of "defensiveness, introspection, and uncertainty." Dillard's words describe the institution fittingly during a time in which the army itself was searching for a singular purpose.[15]

After the war, volunteer veterans who had risen to senior wartime rank questioned regulars' dominance and their implicit claim to ascendent military expertise. Led by former volunteer general John A. Logan, they argued that the Civil War had proven volunteers' mettle alongside professionals, validating America's vaunted citizen-soldier tradition. In a democratic society, they asserted, the army was best led by those who sprang from the citizenry in time of need. In this narrative, West Point—as an academy for would-be generals—stood as a convenient antagonist. But long-serving regulars answered just as adamantly, holding up the regular army as a professional standard, an expert cadre around which the amateurish militia and volunteers would rally in time of national need. Echoing the well-respected William Tecumseh Sherman, they claimed that military leadership in an industrialized world had become the sole realm of educated, trained, and disciplined men who, unlike their volunteer counterparts, did not require months of preparation to achieve minimally tolerable levels of competence in times of war. Modern military leadership, they argued, had grown too complex for those not well steeped in its study and practice. This debate over access to an officer's commission was thus by extension an argument over who might accede to the army's senior ranks—a role that carried significant social authority beyond even the army itself.[16]

Some Republicans believed that if access to West Point and a place in the officer corps was something to be envied and even argued over by white Americans, then that same opportunity should be open to young Black men as well. Reformers quite naturally looked to it as one important element of Black accession to broader social and political spheres following emancipation. If the Civil War's end and the constitutional amendments that followed promised a new beginning for Black men and women in America, then a West Point education and the officer's commission it promised certainly were two clear trappings of having passed through that threshold. The rolls of its graduates reached across American life, from the army's senior ranks to leaders in industry, politics, education, and law. In 1870, a West Point diploma promised more than a lieutenancy in the army; it was a young man's ticket to opportunity.[17]

As a national resource, West Point was by its very design supposed to reflect the citizenry. Candidates received nominations from their congressmen or senators, a system that claimed to ensure a regional balance representative of the nation. But this claim carried with it questions of what that meant in a country still flawed by stark inequality. Access to West Point became a topic of debate during the latter half of the nineteenth century, focusing on the school's academic standards for both entry and graduation. The issue was wrapped within the oft-heard cries of elitism on the one hand and demands for a more rigorous curriculum on the other. Reflecting on admissions generally, the academy's own Board of Visitors, a presidentially appointed body for review and oversight, noted that the "United States Military Academy is not an institution for the benefit of a favored few, nor should it be an experimental arena of the youth of our country" and emphasized that "it belongs to the nation and is supported for the nation's welfare."[18] The academy's standards for entry and academic rigor had to allow the sons of rural Missourians to succeed alongside those of New England's urban families—despite the vastly different schooling that each had benefited from to that point in their lives. With this concurrent debate over access in mind, for Radical Republicans the fact that West Point belonged to the nation translated quite naturally, in the minds of a vocal few, to arguments in favor of Black cadets and eventually Black officers. But for others, this lofty ideal carried the academy's national role to an unacceptable extreme—for them, newly realized Black freedom did not necessitate access.

Across American higher education more generally, very little progress had been made in establishing access and opportunities for Black Americans. Integration remained almost unheard of in the nation's colleges and universities. One historian has remarked that "traditions in education are very strong so that change normally meets with great resistance." If true generally, then this assertion is a pronounced understatement when applied to the period's stark social realities.[19] In a racially defined society, cultural norms were not easily moved by federal acts or public decrees.

Only a small number of progressive white colleges had matriculated Black students at all during the preceding decades. These small steps forward were taken at Antioch, Bowdoin, and Oberlin Colleges and Ohio University, institutions that departed from the day's dominant white conventions and embraced racial advancement as part of their values. For its part, Harvard, already a leader in American education, admitted its first Black student not many years before. At Yale, a Black man graduated from the medical school in 1857 but was the only one of his race to matriculate there for more than a decade. Beyond the overt bigotry that pervaded society, a part of this frustrating pace for progress was the lack of classroom preparation for Black children, most of whom were only recently freed from enslavement. If they had any education at all, it was at best rudimentary except for a rare few. In 1866, just after the war's end, only about ten percent of America's Black population of all ages was literate, let alone prepared to undertake a college curriculum of any kind. The opportunities for advancement through education remained a closed door to the overwhelming majority of freed Black Americans and their children.[20]

By admitting Smith in 1870, West Point was a step ahead of the largest part of contemporary higher education and American society generally. For influential Republicans, it must have seemed a logical extension of Reconstruction's noble mission, institutionalizing the promise of wartime sacrifice. But for others, bent on Smith's failure, it was not nearly so logical or noble an endeavor. Smith and those who followed him at West Point were carefully chosen because their benefactors believed they had the necessary skills to make the attempt with some hope for success, but the challenges they encountered far exceeded those faced by their white counterparts. For West Point's first Black cadets, succeeding academically and surviving socially were challenges compounded by racially fueled hatred that manifested within a spectrum ranging from outright ostracism to physical assault. If this was a test of their preparation to join the army's officer corps and claim the equality extended to them under the law, then the test was relentless and severe.

Race, Politics, and Reconstruction: The First Black Cadets at Old West Point offers a holistic examination of that test. Each contribution stands on its own, but together they also present an essential element of the larger narrative, offering a more complete study of West Point's first experience with integration and the intense political and social pressures that pervaded it. The sad saga of racially fueled American politics is a necessary framework for understanding the experience. Within this milieu, for some, West Point's role was clear—the academy would become a field for what they considered a sacred struggle. For others, the endeavor was a direct challenge to a white supremacist so-cial order—if allowed to succeed, they believed, it would only elevate Black Americans to a position of equality that was not due their race.

By spotlighting the complex interplay between race, politics, and society at the center of Reconstruction-era integration at West Point, this volume also offers a model for interpreting and understanding today's difficult issues that are as much a function of these factors as integration at West Point was a century and a half ago. The 2023 US Supreme Court ruling that curtailed affirmative action admissions to American colleges and universities is a stark reminder that race and its long relationship to the social opportunities held forth by higher education remain unresolved issues in our time. The rul-ing provided an exception for the US service academies, allowing them to continue considering race in admissions. That exception reflects the court's opinion that the academies have a purpose that is distinct from that of their civilian counterparts. Critics of that position filed a federal lawsuit in 2023 to overturn the exception and put an end to consideration of race in service academy admissions.[21]

Race, Politics, and Reconstruction: The First Black Cadets at Old West Point presents a collection of patently historical works. But the issues at its core re-main relevant today, resonating across modern American political, legal, and social discourse. In this narrative of early racial integration at West Point, there may be many important lessons to be learned from the past.

Notes

1. "The West Point Cadets," *New York Times,* June 15, 1877.
2. "West Point," *New York Herald,* June 15, 1867.
3. "Statement Showing the Number of Colored Persons Appointed Candidates for Admission to the U.S. Military Academy," October 21, 1886, Record Group 404, United States Military Academy Library Special Collections and Archives.
4. See, for example, Brian G. Shellum, *Black Cadet in a White Bastion: Charles Young at West Point* (Lincoln: University of Nebraska Press, 2006); and John F. Marzalek, *Assault at West Point: The Court Martial of Johnson Whittaker* (New York: Collier Books, 1994). More recently, Tom Carhart, *Barricades: The First African American West Point Cadets and their Constant Fight for Survival* (Las Vegas, NV: Xlibris, 2020) also falls in this category, as his work is more of a collective biography of the first Black cadets at West Point than an examination of the process of integration.
5. Donald B. Connelly, *John M. Schofield and the Politics of Generalship* (Chapel Hill: University of North Carolina Press, 2006), chap. 11; and Le'Trice D. Donaldson, *Duty beyond the Battlefield* (Carbondale: Southern Illinois University Press, 2020),

chaps. 4–5 stand out. Chapters 4–5 in *Duty beyond the Battlefield* are biographical studies of Henry O. Flipper and Charles Young within a framework established in the first three chapters.

6. See, for example, Theodore J. Crackel, *West Point: A Bicentennial History* (Lawrence: University Press of Kansas, 2002), 145–49; and Edward M. Coffman, *The Old Army: A Portrait of the American Army in Peacetime, 1784–1898* (New York: Oxford University Press, 1986), 225–29.

7. Thomas Fleming, *West Point: The Men and Times of the United States Military Academy* (New York: William Morrow, 1965), 213–31; and Stephen Ambrose, *Duty, Honor, Country: A History of West Point* (Baltimore: Johns Hopkins University Press, 1966), 231–37, have had more sway in the field than they should. Flipper's memoir contains several statements throughout stating that the West Point staff and faculty were perfectly fair and unbiased in their dealings with him. In one fairly representative selection, Flipper writes, "The officers of the institution have never, so far as I can say, shown any prejudice at all. They have treated me with uniform courtesy and impartiality." See Henry O. Flipper, *The Colored Cadet at West Point: Autobiography of Lieut. Henry O. Flipper, U.S.A., First Graduate of Color from the U.S. Military Academy* (New York: Homer Lee & Co., 1878), 122. Too often that characterization has been accepted uncritically by later historians and incorporated into their own work. For example, see William P. Vaughn, "West Point and the First Negro Cadet," *Military Affairs* 35:3 (October 1971): 100–102; Walter Scott Dillard, "The United States Military Academy, 1865–1900: The Uncertain Years" (PhD diss., University of Washington, 1972), chap. 7; and Carhart, *Barricades,* chap. 4.

8. Accordingly, this volume builds on themes central to Donaldson, *Duty beyond the Battlefield;* and Holly A. Pinheiro Jr., *The Families' Civil War: Black Soldiers and the Fight for Racial Justice* (Athens: University of Georgia Press, 2022).

9. Michael W. Fitzgerald, *Splendid Failure: Postwar Reconstruction in the American South* (Chicago: Ivan R. Dee, 2007), 119–21; Eric Foner, *Reconstruction: America's Unfinished Revolution* (New York: Harper & Row, 1988), 230–61. Contestation was not universal, however, as Barbara Gannon has pointed out in *The Won Cause: Black and White Comradeship in the Grand Army of the Republic* (Chapel Hill: University of North Carolina Press, 2011).

10. For treatment of White perspectives on Black officership, see Joseph T. Glatthaar, *Forged in Battle: The Civil War Alliance of Black Soldiers and White Officers* (New York: Free Press, 1990), 28–36, 176–80.

11. Robert Wooster, *The United States Army and the Making of America: From Confederation to Empire, 1775–1903* (Lawrence: University Press of Kansas, 2021), 198–240; Russell F. Weigley, *History of the United States Army,* expanded ed. (Bloomington: Indiana University Press, 1984), 257–70; Robert Wooster, *The American Military Frontiers: The United States Army in the West, 1783–1900* (Albuquerque: University of New Mexico Press, 2009), 163–276; and Michael L. Tate, *The Frontier Army in the Settlement of the American West* (Norman: University of Oklahoma Press, 1999). On

the army's role in mapping the American West, see Richard Bartlett, *Great Surveys of the American* West (Norman: University of Oklahoma Press, 1980); and William Goetzmann, *Army Exploration in the American West: 1803–1863* (Austin: Texas State Historical Society, 1991).

12. Weigley, *History of the United States Army*, 262, 267, 598.

13. Wooster, *US Army and the Making of America*, 222–40; J. P. Clark, *Preparing for War: The Emergence of the Modern U.S. Army, 1815–1917* (Cambridge, MA: Harvard University Press, 2017), 99–135; Weigley, *History of the United States Army*, 267–74.

14. In addition to the many works on the period more generally, Brooks Simpson emphasizes the limits of executive authority in *The Reconstruction Presidents* (Lawrence: University of Kansas Press, 1998).

15. Dillard, "United States Military Academy, 1865–1900," 244. Michael David Cohen places education within the period's social and political backdrop in *Reconstructing the Campus: Higher Education and the American Civil War* (Charlottesville: University of Virginia Press, 2012). On public debate about the proper role of higher education in meritocratic American culture, see Frederick Rudolph's seminal history, *The American College and University: A History* (1962; reprint, with introduction by John R. Thelin, Athens: University of Georgia Press, 1990), 202–20. Other timeless histories of the developing colleges and universities in America are Laurence Vesey's *The Emergence of the American University* (Chicago: University of Chicago Press, 1962); John S. Brubacher and Willis Rudy, *Higher Education in Transition: A History of American Colleges and Universities, 1636–1976*, 3rd ed. (New York: Harper & Row, 1976); and more recently, Christopher Lucas, *American Higher Education: A History* (New York: Palgrave Macmillan, 2006); and John Thelin's updated *A History of American Higher Education*, 2nd ed. (Baltimore: Johns Hopkins University Press, 2011). See also Charles Dorn, *For the Common Good: A New History of Higher Education in America* (Ithaca, NY: Cornell University Press, 2017)—especially the commentary on Black education in the later nineteenth century presented in chap. 8, "The Burden of His Ambition Is to Achieve a Distinguished Career."

16. On the public argument about the regular army's federal role versus that of the amateur tradition, see John A. Logan *The Volunteer Soldier of America* (Chicago: R. S. Peale, 1887); and for secondary histories, Jerry Cooper, *The Rise of the National Guard: The Evolution of the American Militia, 1865–1920* (Lincoln: University of Nebraska Press, 1997). A speech Sherman gave to the Military Service Institute of the United States in 1885 asserted the increasingly complex character of military leadership; see William T. Sherman, "The Militia," *Journal of the Military Service Institute of the United States* (JMSIUS) 6 (March 1885): 1–14. The military's claim to professional status is documented in William Skelton, "Professionalization in the U.S. Army Officer Corps during the Age of Jackson," *Armed Forces and Society* I (Summer 1975): 443–71; William Skelton, *An American Professional of Arms: The Army Officer Corps, 1784–1861* (Lawrence: University of Kansas Press, 1992); and Mark R. Grandstaff, "Preserving the 'Habits and Usages of War': William Tecumseh

Sherman, Professional Reform, and the U.S. Army Officer Corps, 1865–1881, Revisited," *Journal of Military History* 62 (July 1998): 521–46.

17. On the evolving professions in nineteenth-century America as repositories of expertise with social authority, see Burton Bledstein, *Culture of Professionalism: The Middle Class and the Development of Higher Education in America* (New York: W. W. Norton, 1978); and Philip Elliot, *The Sociology of the Professions* (New York: Herder and Herder, 1972).

18. Dillard, "United States Military Academy, 1865–1900," 259. This same debate that pitted admissions standards against accessibility took place across American higher education more generally as well, with the evolving high schools playing an important role: see Rudolph, *American College and University*, 281–86.

19. John D. Pulliam and James Van Patten, *History of American Education,* 6th ed. (Englewood Cliffs, NJ: Prentice-Hall, 1995), 97.

20. Pulliam and Van Patten, *History of American Education,* 98. On Harvard's accommodation of the period's racism, see Jennings L. Wagoner Jr, "The American Compromise: Charles W. Eliot, Black Education, and the New South," in *The History of Higher Education,* eds. Lester F. Goodchild and Harold S. Weschshler, 2nd ed. (Needham Heights, MA: Simon & Schuster, 1997).

21. *Students for Fair Admissions, Inc., v. President and Fellows of Harvard College,* 20-1199, June 29, 2023. This was an opinion written also in response to *Students for Fair Admissions, Inc., v. University of North Carolina, Chapel Hill et al.,* 21–707. The exception for the US Service Academies is found on page 22, note 4 of the court's majority opinion: "The United States as amicus curiae contends that race-based admissions programs further compelling interests at our Nation's military academies. No military academy is a party to these cases, however, and none of the courts below addressed the propriety of race-based admissions systems in that context. This opinion also does not address the issue, in light of the potentially distinct interests that military academies may present." Just a few months after the court's ruling, on September 19, 2023, the same organization that sued Harvard filed a federal suit seeking to end West Point's consideration of race in admissions. See Michael Hill, "West Point Sued over Using Race as an Admissions Factor in the Wake of Landmark Supreme Court Ruling," Associated Press, September 19, 2023.

A NURSERY OF
TREASON REMADE?

Reconstruction Politics and the Rise and
Fall of West Point's First Black Cadets

Adam H. Domby

The Civil War left its stain on innumerable facets of American life, from the families ripped apart by conflicting loyalties to the political turmoil of the presidential assassination. While West Point—the US Military Academy—was not the war's most prominent casualty, it was not spared. Indeed, its reputation was left tarnished by the war—specifically, the number of its graduates who fought for the Confederacy led many Americans to view the academy with suspicion. In 1866, the *Bedford Inquirer* sarcastically noted that instead of political geography, cadets should have to pass a "political morality" exam, "considering the views held by some graduates of that institution on the obligation of oaths."[1] In the years after the war, West Point frequently continued to be critiqued for having been a hotbed of disloyalty.

During political fights over how to shrink and reorganize the army after the war, the topic of West Point and the number of its graduates who were Confederate leaders frequently arose.[2] When proposals were brought forward in 1866 to permanently enlarge the army, those opposed argued that prewar officers "did as much to help the South as the North." Critics not only complained about the number of soldiers who helped the Confederacy but also about the quality of the commanders West Point produced. The *New-York Tribune* noted, "McClellan and Buell and Porter embodied this West Point regular Army spirit" as officers who "did not want to fight," and if they had to fight, wanted the fight to be "perfectly gentlemanly."[3] Fitz John Porter had been cashiered from the army in 1863 for his failures at the Second Battle of Bull Run; Don Carlos Buell and George McClellan were both removed from command in 1862 for being too cautious. These critiques echoed wartime complaints that West Point had been "intensely pro-slavery," "essentially aristocratic," and "generally ha[d] no heart in the war."[4]

While some insisted West Point had been a "nursery of treason," others pointed to William T. Sherman, George Meade, George H. Thomas, and Ulysses S. Grant as examples of how West Point trained the United States' best generals.[5] One newspaper article argued that of 820 West Point graduates, reportedly only 197 joined the Confederacy.[6] Scholars have also debated about how disloyal West Point really was. One scholar found that at least 275 graduates of the academy served in the Confederate military (161 of whom were active duty in the US Army in 1860).[7] West Point's own website states 304 graduates broke their oath to the United States.[8] Not all southern graduates joined the Confederacy. As Ty Seidule has pointed out, seven of eight US Army colonels from Virginia remained with the Union. Only one, Robert E. Lee, "resigned to fight against the United States" thus committing treason as defined by the US Constitution.[9] But as is usually the case, the complex reality of West Point's track record was often irrelevant in political debates. Even Grant's ties to West Point were often held up as evidence that he might be too conservative. In 1868, Charles Moss "feared Grant's West Point sympathies would lead him into the embrace of the Conservative Republicans."[10] Indeed, Moss felt that West Point graduates as a "class of men, as a rule, are conservatives and aristocratic in their views."[11]

In the aftermath of the Civil War, debates about how to view West Point reflected larger conversations about to how to recall the war. After fighting ceased, the nation was left to make sense of what had occurred. Reconstruction, as David Blight has asserted, "was one long referendum on the meaning and memory of the verdict at Appomattox."[12] Indeed, while slavery was dead, what replaced it remained unclear. What did being "free" from slavery entail? Did freedom mean that the previously enslaved would own the fruits of their labor, or was there something more? Would African Americans be given equal rights before the law? Treated equally socially? Would those who had been formerly enslaved get the vote? Who was allowed to hold office? Were Black Americans full citizens? Could they be nominated to West Point? These remained unanswered questions in 1865, and they would be central to the political fights of Reconstruction. While scholars have studied these fights for inclusion within state legislatures, at state constitutional conventions, and in electoral politics in the South, inclusion at West Point was also a significant battleground over the meaning of the war. Throughout the late nineteenth century, each time a Black cadet was appointed to West Point, he became part of a larger political debate around Black rights and abilities, and about equality more generally.

From 1870 to 1889, at least twenty-seven African Americans were nominated to West Point; twelve became cadets, and three graduated.[13] These efforts to open the academy to Black Americans paralleled America's first experiment at opening the ballot with universal manhood suffrage. But just as the gains made by African Americans during Reconstruction were overturned and African American men were disenfranchised, their foothold at West Point was ultimately lost as well. The loss of the ballot during Jim Crow was not just a loss of political power but the loss of a crucial route for Black Americans to serve as officers in the US Army. As W. E. B. Du Bois famously said of Reconstruction, "The slave went free; stood a brief moment in the sun; then moved back again toward slavery."[14] And the same might be said about African American cadets at West Point, as the rise and fall of pathbreaking Black cadets paralleled the fights of Reconstruction.

In 1865, the Thirteenth Amendment made clear slavery was dead, with the notable and important exception of convict leasing. But in 1865 and 1866, during what is often called Presidential Reconstruction, white southerners began forming new governments and passing new state constitutions without the input of their Black neighbors. These new constitutions aimed to limit African American gains to the bare minimum former Confederates thought they could get away with. The new state governments passed "Black Codes," laws that denied full citizenship to African Americans. Bans on Black gun ownership and laws targeting vagrancy, combined with rules about apprenticeship and unequal punishments for the same crime, were just a few of the features of such codes.[15]

In contrast, Black Americans sought to expand their freedoms as much as possible. African Americans in the South laid claim to citizenship, often with arguments that they had earned their rights through military service during the war.[16] The *New Orleans Tribune,* a Black paper, reprinted an article from the *New York Herald* arguing for Black suffrage. The article stated: "Against this demand that as slavery is abolished and that as the African race have powerfully assisted us in putting down the rebellion and in saving the life of the nation, they should have a share in the political right of the ballot-box, what valid objection can be made? We cannot long resist the demand in view of the extinction of slavery and the services of the Southern blacks during this war. . . . We have had two hundred and fifty thousand of them in the service of the army and navy."[17]

The stakes were obvious to many: without enfranchisement, slavery would be replaced with a new system of laws, social norms, and violence that provided disparate rights and oppressed African Americans. The goal of former Confederates was clearly to make "such laws as will allow the freedmen only to be 'hewers of wood and drawers of water.'"[18] Such an outcome also would give control of national politics to the Democratic Party, as the South would not be able to elect Republicans without Black votes. But former Confederates miscalculated. While Andrew Johnson was willing to let Black Codes go into effect, the Republican-dominated Congress was not. When Congress returned to session, they rejected the congressmen that southern states sent—many of them former Confederates—and began work on a new plan for Reconstruction.

While what followed is often depicted as being run by carpetbaggers and northern politicians, many anti-Confederate southern whites, as well as most African Americans, pushed for a more radical Reconstruction. Far from northern imposition, many southerners, both white and Black, wanted to remake society and southern politics. Indeed, those pushing for the harshest Reconstruction measures were frequently white southerners who had been persecuted by the Confederacy.[19] Spurred on by Black southerners and white Unionists in the South who sought relief and aid, Congress attempted to remake southern politics and instituted the Reconstruction Acts. In March 1867, Congress passed the first such act, which put most southern states under military control and laid the blueprints for how southern states could reenter the union. The army was tasked to keep the peace until civil government was restored. To restore civil government, states had to hold elections for delegates who would attend a convention and draft a new constitution. Additionally, these conventions would have to ratify the Fourteenth Amendment, providing equal rights to African American men and thus redefining who was a citizen. These elections had to include African American men as voters. While some white Republicans in the South were hesitant to enfranchise the formerly enslaved, the Republican Party was unlikely to survive in the South without Black voters. While the period that followed is often called "congressional Reconstruction," "military Reconstruction," or "radical Reconstruction," it might best be called biracial Reconstruction to describe those who shaped it within the South.[20]

The Republican Party that came to power across much of the South included over 1,500 Black elected officials over the next decade.[21] White and Black Republican officials relied on the votes of African Americans and white allies—many of whom had been anti-Confederate dissenters during the war—to get elected. In most southern states, Black votes alone were not enough to control the state, and in no state were white Republicans numerous enough to maintain control of state politics. Republicans relied on a fragile biracial alliance and military protection from terrorist violence. Enfranchisement and this biracial alliance in the South meant African Americans now had the ability to be nominated to West Point by a congressman. Thus, the fights for enfranchisement were not just about the vote. African Americans were fighting for economic, social, and educational opportunities as well, to name just a few venues of struggle. Among those multitude of opportunities previously open only for whites that African Americans demanded access to were government appointments.

While most Republican officials were white, conservatives attempted to convince white voters that they faced oppression at the hands of a Republican party controlled by Black Republicans. As it became clear that African Americans would have the vote, conservatives across the country used West Point as an example of supposed "negro domination" in the South to rally support against Reconstruction. In 1868, Democratic papers argued that "'darkey' domination at the south means negro equality, if not superiority, at the north." One paper claimed, "Negro suffrage involves the election of negro congressmen . . . [and] negro congressman will appoint negro cadets to West point; negro cadets will become negro lieutenants in the regular army; they will command, as superiors, white soldiers." Attempting to panic readers, the paper warned, "This is only an entering wedge; social equality comes next."[22] Two years before the first Black cadet even entered the school, admission to West Point had already become part of the broader contest over what roles that Black Americans might have in a reconstructed country.

Fear of Black cadets led some to attempt to ban Black entry into the US Military Academy. In February 1868—before any Black cadets even been appointed—a Wisconsin congressman attempted to amend an appropriations bill funding West Point to include a rider that "no part of the money appropriated by this act shall be paid or applied to the pay or subsistence of any but white cadets." When the amendment failed, many presumed Black cadets would soon arrive on the bluffs overlooking the Hudson.[23]

Meanwhile, African Americans pushed for Black cadets as part of a larger effort seeking the rights and opportunities that came with full citizenship. With African Americans now voters, politicians had to be responsive to them. Among their demands was a share of government jobs and appointments. Indeed, in 1869, African Americans and their allies complained that Grant failed to appoint any Black cadets among his presidential nominees to West Point.[24]

It was in this context that African Americans were first appointed to West Point, though not without a few early setbacks.[25] Frank Smith of South Carolina reportedly requested a cadetship in 1867, but little is known about him.[26] Next, Benjamin Butler attempted to appoint Charles Sumner Wilson, the son of a fallen member of the 55th Massachusetts Volunteers. Butler's attempt to tie the nomination directly to Black military service during the war was no accident. As addressed by Cameron McCoy's essay in this volume, African Americans' wartime service in the United States' armies strengthened postwar claims for Black equality, including the appointment to West Point's corps of cadets. Butler had helped move the cause of freedom through his use of the Confiscation Act to justify freeing enslaved men who reached Fort Monroe (at the mouth of the James River) in 1861. He had commanded Black troops. He would become the chairman of the House Committee on Reconstruction after the war pushing for laws that protected African Americans in the South. Nominating a Black cadet fit with his larger political project of pushing for an effective Reconstruction.[27]

Military service and sacrifice by Black soldiers during the war was cited not only as proof of their manhood but also as having earned African Americans opportunities. The United States Colored Troops and other African American units had clearly shown Black men could soldier and could succeed at the academy. Additionally, an implicit argument underlay the touting of Wilson's father's death in the service—that Black men had earned a spot at West Point by their sacrifice. As when arguing for enfranchisement, Black military sacrifice proved a potent symbol in debates about West Point.[28] While Wilson's father was reported as having died in battle, in fact he had died of pneumonia in a regimental hospital in South Carolina.[29] However, it turned out Wilson was too young to be admitted to the academy, and he would instead end up attending Amherst and Tufts.[30]

Shortly thereafter, success came with the appointment of James Webster Smith. While Smith is often presented as having been formerly enslaved, the truth is more complex.[31] His mother was a free woman of color, and his father was formerly enslaved but apparently free by 1860.[32] Smith was born free and can be found in the 1860 census listed as a free person of color.[33] Later essays in this volume will discuss Smith more fully, but before he could become a cadet, he had to be nominated by a congressman; for that, congressional Reconstruction had to be underway. The fact that Smith's father was involved in Republican politics and even became an alderman for Columbia after the war may have led to James's introduction to Congressman Solomon L. Hoge.[34]

Solomon Hoge had enlisted in the US Army in 1861 and was wounded at Second Bull Run. Unable to continue field duty, but having been educated before the war in law, he served in the Veteran Reserve Corps overseeing courts-martial.[35] After the war he entered the army as an officer intending to make a career of military life. He was appointed acting inspector general of the Freedmen's Bureau for South Carolina, but he resigned from that position and the military in 1868 after being elected as an associate justice to the South Carolina Supreme Court.[36] Though he had practiced law since the 1850s, his credentials were repeatedly questioned by conservatives out to delegitimize him as an unscrupulous carpetbagger. (Hoge was originally from Ohio.)[37] Despite the insults, and having shed blood for the Union, Hoge now sought to protect African Americans in the South.

Hoge was a staunch advocate for Reconstruction. In 1868, when Klan violence swept South Carolina, Hoge refused to give up the dream of creating a biracial democracy in that state.[38] That year the governor sent Hoge to meet with the US secretary of war to inform him of the violence being enacted across South Carolina. At their meeting on October 22, Hoge sought to convince the secretary, John Schofield, to use army troops to reassert martial law by telling him of the rampant racial and political violence.[39] Indeed, Hoge had been with Benjamin Randolph, an African American member of the South Carolina Senate, when Randolph was murdered by the Klan just a week earlier. Hoge had himself narrowly avoided death.[40]

Hoge's election to the US House of Representatives in 1868 was contested by Democrats, who claimed he hadn't won; the Republican majority sat him anyway in 1869.[41] Among the evidence submitted in the investigation into the contested election was the use of violence and armed men at polling places to keep Black men from voting in support for Hoge, as well as ballot-box stuffing.[42] Thus, when Hoge appointed James Webster Smith in 1870, he knew full well that Smith would face resistance. Hoge appointed him with the intent of shaking things up.[43]

Internal Republican politics also likely shaped the nomination. Hoge may have desired to push back against internal party criticisms that too many Republican officeholders were not Black.[44] Indeed, shortly after appointing Smith, he lost his effort at renomination to Robert Elliott, who was Black. Democratic papers attributed this to Hoge being white and Black South Carolinians wanting a Black candidate.[45] Instead, Hoge ran for comptroller general, while Elliot would go on to appoint the first Black cadet to the US Naval Academy.[46]

Smith's nomination was not the most politically controversial one from South Carolina in 1870. In February, just prior to his nomination, Benjamin F. Whittemore (representing the second district of South Carolina) resigned from the House of Representatives when it was discovered he had sold nominations to the nation's military academies.[47] Hoge supported his expulsion from Congress.[48] Whittemore wasn't alone. John DeWeese of North Carolina resigned for the same offence, and Roderick Butler of Tennessee was censured (but not expelled) by the House.[49] These and other high-profile cases of corruption were used by those opposed to Reconstruction to paint Republicans as more corrupt than their opponents—an unfair stigma that the Dunning School would write into history books in the early twentieth century. For example, Congressman Jacob Golladay of Kentucky—a Democrat—was also forced to resign for selling nominations.[50] There were rumors that an Ohio Democrat might have done the same, although nothing seems to have come of an investigation into it.[51] Whittemore's seat would next be held by Joseph Rainey, the first Black person to serve in the US House of Representatives.

This series of scandals may have also helped spur Smith's nomination. Shortly before Hoge made Smith's appointment, Hoge had himself been accused of corruption regarding a nomination to the Naval Academy. The congressman was cleared when it turned out that the father of Hoge's stated second choice paid his first choice to decline an appointment, freeing up a spot for his son. Hoge had no idea about the payments and received none of the money.[52] Still, the taint of scandal may have helped spur him to appoint an African American from his own district to help erase any stain of perceived corruption.

Reform efforts were not just coming from the aftermath of the war but also from the long history of West Point functioning as a finishing school for elites who bought their children entry. Some Americans hoped the fallout from these scandals would make it easier for "poor boys of intelligence" to have a chance at getting appointments to West Point. If the sale of appointments ended, it would mean places at the academy would be "within the reach of any citizen of the United States."[53] This was exactly what Smith's appointment was supposed to prove—not only that any citizen might go to West Point but also that Black men were citizens of equal standing and thus were also entitled to entry.[54] In the midst of these scandals, Hiram Revels joined the US Senate in February 1870, becoming the first African American senator.

Smith was not alone when he arrived in New York. Initially, he roomed with Michael Howard, a Black man appointed from Mississippi. Though appointed to West Point, Smith and Howard were not yet cadets. A medical board nearly prevented Smith from entering: claims that Smith's eyesight was poor were put forward but failed to stop him becoming a cadet, and he was accepted on a probationary status. A second medical board reviewed his case a year later and found any short-sightedness was not a problem.[55] Howard, on the other hand, failed to pass the academic entrance exams.[56] While those opposed to Black progress criticized Smith's becoming a cadet, others defended his advancement. The *Yorkville Enquirer* noted that those white cadets complaining about his presence should recall that their own appointment to West Point "was achieved solely by the efforts of their political friends."[57]

In newspapers around the country the arrival of Smith and Howard at West Point continued to be tied to biracial Reconstruction politics. In the run-up to their examinations, Democratic papers tried to paint the appointment of

Black cadets as a cheap appeal toward Black votes. Conservatives claimed Republican congressman supposedly knew Smith, Wilson, and Howard would never be able to pass the necessary examinations. Accusations that Butler had known his cadet was too young and "only appointed him for political purposes" abounded. One Georgia paper claimed these appointments were "mete buncombe, designed to retain the negro vote," as the congressmen knew "their appointments would not be confirmed." They even fabricated claims that Smith was nearly blind and had bad lungs, so he would surely fail to pass the physical requirements. Such attacks served not only to undermine Republican politicians but also to undermine the Black freedom struggle by implying African Americans were incapable of entering West Point, let alone graduating. Democrats declared that the supposed inferiority of Black men would keep them out, claiming, "five hundred may be appointed but the dreadful ogre, 'incapacity,' will slaughter them all."[58]

African Americans were also concerned that politicians were just appointing Black cadets for cheap political points. After Wilson was rejected for his age and Howard for his literacy, and with reports that Smith was likely to fail due to his eyesight, the *San Francisco Elevator* questioned if "members of Congress know the necessary age and qualifications? or do they set up colored boys like ten-pins to be knocked down? We hope none will be nominated in future for sham."[59] This was not to say that African Americans did not want to see more appointments. Quite the opposite: they celebrated appointments, wanted cadets to be successful, and recognized the stakes. Le'Trice Donaldson's essay in this volume discusses more fully how African Americans responded to Black cadets.

Efforts to get African Americans into West Point in 1870 were clearly tied to the larger struggles to gain equality and opportunities in a reconstructed South. Both those in favor of and those opposed to Black rights recognized these fights were related. Indeed, in June 1870, those decrying their appointments referred to Smith and Howard as "the Fifteenth Amendments," a reference to the constitutional amendment protecting African American voting rights, which had been ratified earlier that year.[60] The *National Anti-Slavery Standard* compared the struggles of Black cadets to efforts to desegregate streetcars and steamboats as well as to those who had opposed slavery, noting that "slavery in its spirit and essence, cannot truly be said to have been abolished while negro hatred which grew out of it continues so rife."[61] West Point remained an important political battleground in the years after the war, and the fight over Black cadets was part of a series of fights over what emancipation really meant and whether African Americans would be treated as full citizens. These fights were interconnected. It was clear to many observers that without Black enfranchisement, there was little chance of Black cadets assuming a place at West Point.[62]

In addition to the racial politics of Reconstruction and anti-elitist sentiment, the legacy of West Point graduates' betrayal of the Union during the war continued to shape debates around reform and West Point. When Andrew Johnson issued his May 29, 1865, proclamation of amnesty for former Confederates, he explicitly excluded from amnesty "all military and naval officers in the rebel service who were educated by the Government in the Military Academy at West Point, or the United States Naval Academy."[63] When the Amnesty Act passed in 1872, most former Confederates saw any "political disabilities" removed. However, among those who remained without amnesty—alongside former elected officials—were former military officers who fought for the Confederacy. Thus, West Point graduates who served in Confederate forces remained disenfranchised. In debates around amnesty, West Point and Annapolis graduates who sided with the Confederacy remained a contentious issue. Among those opposed to premature amnesty was Congressman Hoge. In early 1871 Hoge stated on the House floor that while he might support amnesty in the future, "it would not be while the people of his district were shot and murdered and taken out of their houses at night, and whipped and otherwise maltreated. He was opposed to this forcing amnesty upon men who declared they did not desire it."[64] Hoge remained committed to Reconstruction, ensuring African Americans retained the ballot, and to integrating West Point. He would appoint another African American cadet in 1876.

Hoge was not alone in seeking to integrate West Point. Henry Alonzo Napier of Tennessee would soon follow Smith, appointed by Congressman William F. Prosser in 1872. Prosser had announced in 1870 that he was seeking a Black man who could pass the examination and enter the school.[65] Democratic papers continued to try to present nominations of Black cadets as a way for Republicans to "cover up their rascalities," with claims that Prosser had been part of the nominations-for-money scandal. Whether he was trying to protect his own reputation with "the negro cadet nomination dodge" or genuinely wanted to appoint a Black cadet, Prosser recognized the importance of his nominee being successful.[66] Each failure would be held up as an example of Black inferiority. In an example of what Ibram X. Kendi calls "uplift suasion," many supporters of Black cadets hoped their success would prove to onlookers that African Americans were deserving of equal treatment.[67] Instead, as is all too often the case, conservatives opposed to Black progress claimed that any success was the exception that proved the rule, deriding Black firsts as the result of an atypical exemplary individual, unrepresentative of his or her race.

As Reconstruction continued, those supporting Black cadets drew on wartime history to justify additional appointments. The *South Kansas Tribune,* for example, pointed out in 1871 that "the people have but a low estimate of the West Point business at best. Two thirds of those what were educated there were rebel officers and it has been little else than a nursery for treason." The addition of Black men, then, who "exhibited many soldierly qualities, and among them the best of them were loyalty, patience and devotion," remained a welcome change to those still angry over the war. The *Tribune* celebrated that in the future, it was "not likely that colored cadets would prove untrue to the Government that has disenthralled, and now educates, elevates, and respects them." Thus, enrolling Black cadets served not only to prove Black citizenship, demonstrate Black ability, and provide opportunities to Black men, but also to help redeem West Point from its reputation of being open to only "the sons of northern aristocrats and southern nabobs." Indeed, for some in Kansas, the presence of a Mormon—and son of Brigham Young—was far more concerning than was the arrival of Napier. The paper noted "it does not seem that the Government ought to educate soldiers for mormondom."[68] Those supporting integration included many who remembered the disloyalty of southern graduates during the war. Frederick Douglass's *New National Era* noted in 1874 that "the South once ruled West Point, much to its detriment in loyalty."[69]

Not all appointments of Black cadets came from radical Republicans. Moderate Republican James C. Freeman of Georgia, whose district was 49 percent Black, appointed the first Black graduate of West Point. Freeman had been a large slaveholder before the war but opposed secession.[70] He nominated Henry O. Flipper. Flipper—who will be discussed more fully later in this book—would be the first to graduate.

Appointing a Black cadet was not without risk to congressmen. Always looking to paint Republicans as corrupt, Freeman was accused of not announcing a "vacancy" at West Point, which was why no "white folks" supposedly applied for the spot. Ignoring the fact that many white congressmen appointed their friends and supporters, Democrats saw this nomination as a useful tool with which to hammer him, claiming they "disapprove in *toto* of his appointment of the negro Flipper to West Point, where there were so many young white men who would gladly have accepted the place, not so much because Flipper is a negro, but because it is carrying out the policy of extremists who want to *force* the negro over the white man." Given the nature of his district, Freeman had to appeal to both white and Black voters. He claimed he offered the appointment to a white man first, but Democrats seeking to defeat Freeman would not have been satisfied with anything that resulted in Black success. Freeman was constantly attacked for wanting "social equality."[71] Conservatives clearly saw this as an opportunity to paint Freeman as someone who wanted not just equality but Black domination of white southerners. Attacking Republicans for covering up corruption and making cheap appeals to Black voters with nominations bound to fail must have created cognitive dissonance when paired with claims that these same nominations aimed to put Blacks over whites, but logic has rarely stood in the way of white supremacist propaganda.

Not all Black nominees were nominated by white congressmen. Josiah Walls of Florida would appoint the fourth Black cadet. Walls had been enslaved in Virginia before the war. Like Hoge, he would have his election repeatedly challenged by conservatives unwilling to accept defeat. A staunch advocate for public and compulsory education, Walls unsurprisingly saw integrating West Point as valuable.[72] In 1872, Walls nominated Thomas V. R. Gibbs, the son of Jonathon Clarkson Gibbs (a fellow Black politician in Florida). Perhaps trying to temper complaints that he was only helping Black Floridians, Walls also nominated the son of former Florida governor and conservative politician David Walker, who had overseen the institution of Black Codes after the war and opposed the 1868 constitution.[73] Instead, Walls was criticized for it.[74] Indeed, locally at least, his appointment of Walker may have been a bigger controversy than his appointment of a Black cadet.

As with other cadets, Gibbs in many ways became a proxy for debates about African American intelligence and ability. Papers opposed to his success and to African American progress portrayed him as awkward, claiming that "his feet and hands are out of proportions to his form" and that he was smaller than other cadets. Discussions around the purity of his blood also appeared (as they had with lighter-skinned Smith and darker-skinned Howard in 1870).[75] Gibbs was presented as unclean, unintelligent, and unable to master skills, and, like others, his eventual failure to graduate was held up as evidence that Black men did not belong at West Point.

Nor was every appointment made by a southern Republican needing Black votes—quite the opposite in the case of Charles A. Minnie, who was appointed by a member of Tammany Hall.[76] In 1877, Staten Island Democratic congressman Nicolas Muller found himself "bothered a good deal by the totally unexpected result of the competition for the West Point cadetship in his district." Offering an examination for those wanting an appointment allowed Muller to avoid accusations of patronage, but "it never occurred to him for a moment that the prize might be carried off by a colored boy." While the results angered some white constituents, Black New Yorkers were excited by the prospect of Minnie's success, seeing it as a sign of progress.[77] Minnie, however, stayed at West Point only a year, finding the social ostracization and mistreatment too much to endure.[78]

It is important to note that appointments to West Point were always political and still are. Even those who used examinations to determine which local candidate they should appoint would trumpet their use of an exam to demonstrate how uncorrupt they were. Still, the majority of nineteenth-century Black nominees (twenty-one of twenty-seven) came from the South, and Black members of Congress nominated eleven of them. Five were nominated by Congressman Robert Smalls alone.[79]

Politics shaped the experiences of Black cadets as well. As discussed in the following essays, all of the early Black cadets faced ostracization, insults, and mistreatment. Among the few white cadets who treated Flipper better were reportedly Ben Butler's son and John Bigelow, whose father had been the American minister to France under Lincoln.[80] On the other hand, President Grant's failure to take a more active role in protecting Black cadets may be related to the fact that the president's son was a cadet at the time and may have been involved in the harassment of Smith.[81]

Those who hoped Black success at West Point might lead white Americans to change their racist views were sadly disappointed. Flipper's graduation did not make it any easier for subsequent African American cadets or lead many of his skeptics to question their past doubts. Despite Flipper's seeming success, many white southerners still questioned his abilities and expected he would fail as a soldier. Falling back on race "science," the *Austin American-Statesman* noted, "West Point . . . will hardly change the shape or thickness of the Flipper Skull or inherent qualities of the Flipper brain power," while dismissing his success as the product of being "an exceptional African."[82] Excusing his success and presenting Flipper as an anomaly both were often couched in naysayers' claims that they did not wish him failure, but they expected it because of his race and because white officers would never assent to treat him as a peer.[83] Raising the bar that African American cadets must hurdle similarly served to ensure they would not be judged successful. First, papers decried Flipper's ability to get in, then to graduate; and when Flipper graduated, they moved to noting he could not be accepted socially by white officers. Changing of the meaning of success ensured that those wishing to find failure could declare "The Enactment of Equality—A Failure," no matter how much African Americans succeeded.[84] As for Flipper, he rejected "African Americans' notions of collective freedom and masculine identity," never having embraced "the notion of helping the entirety of the race" through his accomplishments. Instead, he saw his service as not "belonging to anyone but himself."[85]

Flipper graduated West Point and entered the army just as Reconstruction was being dismantled. In 1874, Democrats took control of the House of Representatives, and two years later the 1876 presidential election was contested. A compromise in 1877 resulted in Rutherford B. Hayes becoming president and the end of military support for the remaining biracial Reconstruction governments in the former Confederacy. Without this support, former Confederates could use violence, intimidation, and fraud—and, eventually, legal means like literacy requirements—to disenfranchise African Americans across the South. A counterrevolution against the changes Reconstruction brought seemed to be overturning the progress made after the war.

In 1876, Democrats in the House attempted to pass an act allowing former Confederates to be appointed to the US Army. The worst part about such a law, the *Pittsburgh Commercial* noted, was that it would "allow southern ex-rebels to be appointed to professorships at West Point, so as to poison our army at its fountain, by teaching the cadets who are to be the future officers of our army the Southern disloyalty doctrines."[86] With former Confederates as professors, Black cadets would have been doomed to failure. The proposal failed primarily because of the Republican-controlled Senate.

There would be more nominations of Black cadets even after the fall of the last Reconstruction-era Republican state government in the South. Reconstruction, as a process, did not just end in 1876 and Jim Crow begin the next day. By the late 1870s it had been fading over time. Some states saw their biracial Republican-controlled governments fall in 1870, while others did not collapse until 1876. Similarly, the creation of the Jim Crow order was contested and took time. In 1876, it was not clear that Republicans would not regain control of some southern states in the next election cycle. Indeed, in both Virginia and North Carolina, Republicans regained a semblance of power in the decades to come with the Readjuster and Fusion movements. African American power waned over time before legal disfranchisement ended Black electoral political power in the South almost entirely. In fact, Thomas Gibbs, the third African American to attend West Point, was elected to the Florida House of Representatives in 1885 and 1887.[87] He also served in Florida's 1885 state constitutional convention; the constitution that was drafted, which made poll taxes legal and mandated segregated schools, was in many ways a reaction to Reconstruction. It sought to undo the progress made in the 1868 biracial convention. As well as fighting against poll taxes and school segregation, Gibbs sought to have a normal school and an agricultural school established for Black Floridians.[88]

Put simply, African American efforts for freedom did not end in 1876, hibernate for seventy-eight years, and reappear in 1954. Black Americans continued fighting for their civil rights despite setbacks (and continue to do so today).[89] While Reconstruction is often depicted as ending with the compromise of 1877, it was not until 1878 that Republicans lost the US Senate.[90] Indeed, Blanche Bruce, the second Black senator (representing Mississippi), remained in office until 1881, and a few African American representatives held office until 1887. Even then, Black electoral power remained, albeit reduced, and five more Black congressmen would serve in the 1890s.

However, Mississippi's 1890 constitutional convention "changed everything," providing a model for other states to copy as they firmly established Jim Crow rule by legally disenfranchising African Americans through literacy exams and poll taxes. While on the surface these features were "racially neutral," they clearly targeted Black voters by circumventing the Fourteenth and Fifteenth amendments.[91] Other states soon followed suit.[92] Still, it was not until 1901, when North Carolina's grandfather clause came into effect, that the last Black congressman of the nineteenth century left his seat, ushering in a seventy-year period without a southern Black representative in Congress.[93]

The rise and fall of African Americans at West Point in the nineteenth century paralleled the rise and fall of Black electoral power in the American South. Black Americans kept applying, and a few attended West Point until 1889. As Jim Crow rose, the loss of political power by Black Americans ended their ability to get appointed to West Point. The use of congressional appointments had long ensured that only those with political sway had the ability to attend the military academy, so it was hardly surprising that with the loss of power, so went access to academy nominations. The ties between the vote and access to nominations were not new. In 1870, the *Yorkville Enquirer* had noted that "in the light of the Fifteenth amendment, what we shall do with the African in our national and naval academies, is a grave question for the unregenerated mind."[94] Twenty years later, with a lack of Black voters in the Jim Crow South, what did a white southern politician gain by nominating a Black cadet?

It was not just white politicians who sought to keep Black men out of West Point. In April 1880 Cadet Johnson Whittaker was brutally assaulted by three white cadets as he slept. Battered with a club and fists, he was pummeled and choked by white cadets before being tied to his bedframe on the floor. Declaring they would "mark him like they do the hogs down South" they cut the lobe off his left ear and slashed his right ear in multiple places. He was found the next morning "in a half conscious condition."[95] The assault on Cadet Johnson Whittaker by white cadets—discussed in Makonen Campbell and Louisa Koebrich's essay in this volume—was a national news story in 1880.

Each Black cadet—whether he wanted to or not—became part of national political fights around Black rights, abilities, and equality more generally. As Philip Dray has explained, the academy became "a kind of proving ground for questions of equal treatment of the races."[96] Whittaker, who like Smith had been appointed by Hoge, was eventually found guilty of lying about the attack. In an egregious case of blaming the victim, it was Whittaker who was punished for the assault.

Black newspapers decried the attempt to blame Whittaker, comparing it to similar false claims that "all the outrages on colored men [in the] South [. . . were] radical lies invented for effect on Northern elections."[97] The Republican Convention of South Carolina passed a resolution denouncing the assault.[98] The *Indianapolis Leader* noted that the "diabolical savagery committed upon Cadet Whittaker and through him upon the whole Negro race" was bad enough, but blaming Whittaker made clear that "the Academy is in the hands of the same aristocratic, Negro-hating, secession-sympathizing, unrepublican gang who controlled it previous to the beginning of the late war." The paper called for the national Republican convention to take action.[99]

Black cadets were constantly used by those engaged in debates about race and politics in America. Far from being a successful uplift suasion campaign that might convince whites that Black men could make good officers, the supposed failures of Black cadets were frequently held up as evidence that desegregation and social equality were a foolish goal. The Whittaker attack and investigation led the *Memphis Daily Appeal* to declare, "It has been demonstrated at West Point, as elsewhere, that the Caucasian and African can never be forced into social equality."[100] In 1882, with Whittaker's failure, the *Buffalo Sunday Morning News* examined the careers of the eleven Black cadets, wondering, "Can Caste be Conquered?"[101]

Graduation was not a guarantee of respect, either. Even once Flipper graduated, his career was a political hot potato, and efforts were made to derail it. He remained only a test case in the minds of many. While his success might be seen by some as a success for all African Americans, any setbacks might similarly be presented as evidence of racial inferiority. Charged with both embezzlement and conduct unbecoming of an officer and a gentleman, he was acquitted of the former but convicted of the latter. His defenders claimed the case "was prejudged from the beginning" due to his race.[102] Flipper's eventual court-martial and dishonorable discharge in 1882 (posthumously overturned by Bill Clinton in 1999) led many opponents of Black success to crow. Later that year, when Lemuel Livingston failed the entrance exam at West Point, the *Buffalo Morning Express* celebrated that "the country is saved a deal of discussion and perhaps another scandal by this."[103] Not until John Hanks Alexander in 1887 would another Black man graduate from West Point.

When a Black cadet was successful, those opposed to his success conjured explanations to ensure that his accomplishments were seen not as evidence of Black ability but as aberrations. For example, when John Hanks Alexander graduated in 1887, newspapers played up that he was "almost white" and "not a negro, but a quadroon, so light that no one not previously informed, would suppose him to have any African blood."[104] Two years later, Charles Denton Young would graduate, and the *Macon Telegraph* reported that he "had failed to pass the final examination, but for some reason, was given a chance to make up his deficiency." The paper ignored Alexander's existence, claiming Young was the second Black cadet to graduate, and went on to note Flipper's dismissal from service. Despite nominally being about Charles Young's success, the article concluded that "the coeducation of the races at West Point has been a miserable failure so far."[105]

Those striving to advance Black rights were aware of such claims and forcibly pushed back. The *Philadelphia Times* pointed out in 1876 that Flipper "has the negro features strongly developed, but in color he is rather light." Having also noted that white cadets refused to befriend or treat Flipper with respect, the *Savannah Tribune,* a Black newspaper, caustically replied, "It is a pity that your whole race did not have less to do with the negro, and then this young colored cadet with whom you 'don't have anything to do with when off duty,' would not have been in color so light."[106]

Former Black cadets also countered the anti-Black narrative. After leaving West Point, Whittaker defended those who came before him, arguing that Smith "was not sent out on his merits" and that Napier and McKinley had both graduated from other schools. Whitaker noted that Charles Minnie had been nominated by a Democratic congressman after scoring the highest on a competitive exam but that his mistreatment at West Point led him to leave the academy. Those who attempted to desegregate recognized that their failure was not their fault. Rather, responsibility lay on an institution, a country, a society, and a political system still in need of dramatic reforms.[107]

The contributions to the Confederate war effort by West Point graduates and a historical mistrust dating from the nation's founding of professional armies as antirepublican meant that many Americans already looked at West Point with suspicion. The ill-treatment of Black cadets helped remind the public of West Point's already tainted reputation. While in the early twenty-first century, attending West Point is often held up as a badge of pride by political candidates, in the late nineteenth it could still be used as an attack. In 1880, Winfield Scott Hancock ran for president as a Democrat. His education at West Point was frequently contrasted with that of his opponent, James Garfield, who had enlisted in the US Army during the war but never attended West Point. Republican senator George Edmunds conceded that Hancock was a good general and loyal officer during the war but argued, "Now, which way should we go? For the great soldier who is nothing but a soldier . . . or the equally great General, who is of the people and not of West Point."[108] Republican newspapers claimed that Hancock was "imbued and troubled with the highest degree of West Point arrogance and martinet's self-reliant impudence."[109] West Point's reputation for breeding arrogance stemmed in part from its reputation as a place that had trained southern aristocrats before the Civil War.

Indeed, the treason committed by so many southern graduates was not forgotten twenty years after secession. Even as Hancock was being lambasted for his "West Point arrogance," white cadets who mistreated Black cadets were seen as a product of the same system. Critics feared white cadets might still be gaining "the odor of West Point arrogance" that had imbued the regular army during the Civil War. Calling for reform after the Whittaker scandal, the *Rutland Weekly Herald and Globe* complained that academy leadership had allowed "without a word of rebuke or remonstrance, a colored cadet to be treated with systematic indignity, and condemned to utter social isolation." The paper argued that the superintendent of West Point thus should be removed from office regardless of whether Whittaker had been attacked. Reform was necessary, the paper asserted, because "the government cannot and will not permit West Point to become a mere kennel for military puppies, as it once was for those rebel dogs of war, the deep hate of whose disloyalty was written so often in Northern blood. West Point taught loyalty so doubtfully before the war that we found when the war broke out that Southern graduates were only so many knives sharpened on the national whetstone for the national throat."[110]

The treatment of Black cadets led critics to argue that West Point had not really changed since the Civil War. The *Marysville Appeal* argued after the attack on Whittaker that "West Point has been a hot-bed of aristocracy since its doors were opened, and it will prove of no great national loss if it is abolished." The *Appeal* believed that northern cadets had been "converted to that Southern prejudice" by white southern cadets. In response, one Democrat recycled an old talking point and argued that Republican-appointed white cadets mistreating Black cadets demonstrated that "it is the negro vote that Republicans want—not the negro with it."[111] In Congress, Senator William Allison argued that the president should annually appoint two Black men to West Point. He and others recognized that the small number of Black cadets, combined with their mistreatment, meant that African Americans were "not excluded by law but they are practically, and have been for many years."[112] The proposal failed, but African American cadets clearly remained a potent political topic.

West Point's Civil War reputation as a hotbed of traitors did not disappear quickly. Even in 1890, as Albion Tourgée and Lew Wallace continued to argue for reforms at the academy, they intended to take the army further from its antebellum existence of "Northern mudsills officered by Southern gentleman."

Wallace is perhaps best known as the author of *Ben Hur* and Tourgée as the author of *A Fool's Errand, by One of the Fools.* Tourgée and Wallace were both Civil War veterans and Republican politicians who had served in the US Army but had not attended West Point. They wanted to appoint all cadets from the enlisted ranks—which would have presumably destroyed the last vestiges of West Point's reputation as a finishing school for white elites.[113] Instead, West Point's first attempts at desegregation failed.

In the end, these early efforts to provide equal access to American military service did not lead immediately to lasting change. With the rise of Jim Crow and the decline in the number of Black elected officials and voters, political power for African Americans waned. And with that waning political power, appointments of African Americans to West Point all but ended. From 1889 until 1918 no Black cadets entered the academy. Not until Benjamin O. Davis Jr., in 1936, would another Black cadet graduate from West Point. Like those before him, Davis would also face discrimination from the other cadets.[114] Twenty-five years later, in 1961, there were only eleven Black cadets attending West Point.[115]

As of 2023, 12 percent of cadets are Black or African American, but the campus still struggles to welcome nonwhite cadets.[116] Black graduates describe facing racist comments from their white peers.[117] In 2016, sixteen Black female cadets were investigated for taking a photo of themselves with raised fists. Unsurprisingly, they were cleared of any violation of army policy, but the fact that they were even investigated is a reminder of how much racism still pervades American institutions.[118] Simone Askew, the first African American woman to be named the top cadet of her class (in 2017), faced racist harassment.[119] In 2020, after the superintendent of West Point said the school "does not have a systemic problem with racism," Askew and nine other accomplished Black graduates disagreed and published a policy proposal calling for "an Anti-Racist West Point." Risking career harm, these young officers called out how "systemic racism continues to exist at West Point." In their memo they noted how Black students still experienced racist harassment, including being called the n-word while being spit on and in one case having a noose left on their desk. Echoing the experiences of their nineteenth-century predecessors, students who complained about being called the n-word have been accused of lying, leading to honor code investigations initiated against them instead of the students who racially abused them. Disparate punishments for white and Black students for the same offense have been reported.[120] Only in 2022 has the academy attempted to address racist physical infrastructure on campus, including a barracks named after Robert E. Lee. Yet there remained many who objected to changing the campus in any way that might welcome nonwhite cadets.[121] While changes to campus are ongoing at the time of this writing, it seems reform is still needed at the academy over a century and a half after Smith first arrived on campus.

Although the admission of Black cadets may not seem as openly contested as it once was, there remain elements of American society opposed to attempts to increase the number of Black cadets. During Supreme Court arguments over affirmative action in October 2022, one argument put forward defending affirmative action was the need for a diverse officer corps that represents the nation as a whole and the troops they lead. As the solicitor general noted to the Supreme Court, "Our armed forces know from hard experience that when we do not have a diverse officer corps that is broadly reflective of the diverse fighting force, our strength and cohesion and military readiness suffer."[122] The danger is clear. If affirmative action is banned in college admissions and campuses become less diverse, this would impact the pool of officers from whom the ROTC can recruit as well as potentially change West Point's own admissions policies.

Most of the arguments for diversity were ignored, and in June 2023 the court ruled that the University of North Carolina and Harvard cannot consider race in their admission process. Still, the court drew a unique carveout in their decision while otherwise seeking to dismantle affirmative action in admissions at nonmilitary colleges and universities. A footnote explicitly said that "in light of the potentially distinct interests that military academies may present," the decision does not apply to West Point and other military academies. The court tacitly conceded that diversity may actually strengthen institutions even as they sought to ban efforts to diversify educational institutions.[123]

But the fight is not over. The same group that successfully sued Harvard has filed another suit aimed at barring the consideration of race in admissions at West Point.[124] The court may side with them making this carveout temporary. What will occur is not yet determined as of early 2024. The court might recognize the importance of a diverse army to America's national security, or an increasingly conservative court might dismantle long-held precedent over a series of cases rather than in one decision. The Supreme Court has always been a political body as much as Americans may try and pretend otherwise.[125] With no mechanism to enforce decisions, the court relies on the public and the other branches of government accepting its decisions as legitimate. Not wanting to appear that they are ignoring the principle of *stare decisis*—which might make a decision appear flagrantly partisan or illegitimate—justices have at times chosen to chip away at precedent over a series of rulings and still eventually achieve a long-held political agenda. Along the way, however, they also occasionally respond to pushback over an unpopular decision. According to historian Heather Cox Richardson, public opinion and pressure can lead justices to moderate their "more extreme impulses."[126]

Overturning precedent over a series of decisions and even over a number of years allows justices to gauge what popular opinion will accept while also making their decisions appear less partisan, thus protecting the court's reputation. Justices have historically responded to public outcry and threats to the court's power. In the 1930s, after the court ruled against Roosevelt's New Deal repeatedly, the nation responded. An overwhelming electoral victory for Roosevelt in 1936 and efforts in Congress to expand the court put pressure on the justices. They reversed course on future decisions around New Deal program, upholding them as legal.[127] Is this an example of a case where SCOTUS will slowly overturn a precedent over a series of cases, or will the justices find that national security demands the army maintain a diverse officer corps? Time will tell.

If the overturning of Reconstruction teaches us nothing else, it teaches us that progress is not guaranteed, and what goes forward can go back. Will the progress that West Point has made since 1936 be reversed? If so, how far will the army regress? Or will the sacrifices of those nineteenth-century groundbreakers inspire the academy's leaders to make their campus a better place in the twenty-first century? In the past decade white supremacist views have become increasingly accepted in mainstream politics on the right.[128] With overt challenges to Black progress and appeals to white supremacy increasingly featuring in national political discourse over the past decade, what does the future hold for West Point's Black cadets? More generally, how will racism shape American culture, politics, and the military?[129]

These questions remain to be answered and are of importance to the national security of the United States. At the start of the third decade of the twenty-first century, we have already witnessed repeated efforts to roll back the Voting Rights Act alongside attempts to overturn free and fair elections. With increasing political violence and authoritarian rhetoric in the United States—some of it directly tied to white supremacy—Reconstruction history has found new relevance in recent years; perhaps Reconstruction's legacy still echoes at West Point. It seems the fights over the meaning of freedom, equality, and citizenship that those first Black West Point cadets helped shape from 1870 to 1889 remain very much alive today.

Notes

The author would like to thank Lauren Simpson, Holly Pinheiro, Hilary Green, Jennifer Kosmin, Tommy Sheppard, and Amanda Nagel for their invaluable feedback on this essay.

1. *Bedford Inquirer,* February 9, 1866, 2.

2. See, for example, "The Regular Army," *Buffalo Commercial,* January 25, 1866, 2.

3. "The Regular Army," *New York Tribune,* January 20, 1866, 2.

4. "Our Washington Letter," *Chicago Tribune,* February 1, 1862, 2. Accusations of being aristocratic were not new to West Point, having been a critique of the academy during the antebellum era. See Wayne Wei-Siang Hsieh, *West Pointers and the Civil War: The Old Army in War and Peace* (Chapel Hill: University of North Carolina Press, 2009), 12

5. "Reorganizing the Army," *Philadelphia Inquirer,* April 24, 1866, 2. For another example, see "West Point and the War," *Philadelphia Inquirer,* June 30, 1865, 4.

6. "A Vindication of the Regular Army," *Buffalo Courier,* January 25, 1866, 2.

7. James L. Morrison, "The Struggle between Sectionalism and Nationalism at Ante-Bellum West Point, 1830–1861," *Civil War History* 19. 2 (June 1973): 147. Additionally, there were at least forty who fought for the Union.

8. "A Brief History of West Point," United States Military Academy West Point, https://www.westpoint.edu/about/history-of-west-point.

9. Ty Seidule, *Robert E. Lee and Me: A Southerner's Reckoning with the Myth of the Lost Cause* (New York: St. Martin's, 2021), 223.

10. Charles E. Moss, "Letter from Col. Charles E. Moss," *National Anti-Slavery Standard,* October 3, 1868, 1.

11. Charles E. Moss, "Reconstruction—Gen Grant," *National Anti-Slavery Standard,* January 11, 1868, 1.

12. David W. Blight, *Race and Reunion: The Civil War in American Memory* (Cambridge, MA: Belknap Press of Harvard University Press, 2003), 31; see also David Blight, "Lecture 21—Andrew Johnson and the Radicals: A Contest over the Meaning," Open Yale Courses, https://oyc.yale.edu/history/hist-119/lecture-21.

13. Selected Documents Relating to Blacks Nominated for Appointment to the US Military Academy during the 19th Century, 1870–1887, microfilm M1002, National Archives and Records Administration, https://www.familysearch.org/ark:/61903 /3:1:3QHV-P3N2-599P-S?i=7&cat=354502.

14. W. E. B. Du Bois, *Black Reconstruction in America* (New York: Atheneum, 1992), 30.

15. William C. Harris, "Black Codes," *NCPedia,* 2006, https://www.ncpedia.org/black -codes; Steven E. Nash, *Reconstruction's Ragged Edge: The Politics of Postwar Life in the Southern Mountains* (Chapel Hill: University of North Carolina Press, 2016), 34; Deborah Beckel, *Radical Reform: Interracial Politics in Post-Emancipation North Carolina* (Charlottesville: University of Virginia Press, 2011), 47–48. Beckel is an excellent introduction to North Carolina politics in the nineteenth century.

16. For more on the fight for full citizenship and the role of military service, there is an extensive histography: see Le'Trice D. Donaldson, *Duty beyond the Battlefield: African American Soldiers Fight for Racial Uplift, Citizenship, and Manhood, 1870–1920* (Carbondale: Southern Illinois University Press, 2020); Holly A. Pinheiro Jr, *The Families' Civil War: Black Soldiers and the Fight for Racial Justice* (Athens: University of Georgia Press, 2022); James G. Mendez, *A Great Sacrifice: Northern Black Soldiers, Their Families, and the Experience of Civil War* (New York: Fordham University Press, 2019); Brian Taylor, *Fighting for Citizenship: Black Northerners and*

the Debate over Military Service in the Civil War (Chapel Hill: University of North Carolina Press, 2020); Donald Robert Shaffer, *After the Glory: The Struggles of Black Civil War Veterans* (Lawrence: University Press of Kansas, 2004); these discussions about manhood, citizenship, military service, and race did not end after the Civil War. For later examples, see Amanda Nagel, "Democracy for Whom? The Spanish-American War, the Philippine-American War, World War I, and the NAACP" (PhD diss., University of Mississippi, 2014); Chad L. Williams, *Torchbearers of Democracy: African American Soldiers in the World War I Era* (Chapel Hill: University of North Carolina Press, 2010); Adriane Lentz-Smith, *Freedom Struggles: African Americans and World War I* (Cambridge, MA: Harvard University Press, 2011); Kimberley Phillips Boehm, *War! What Is It Good For? Black Freedom Struggles and the U.S. Military from World War II to Iraq* (Chapel Hill: University of North Carolina Press, 2012).

17. "Additional from the North," *New Orleans Tribune,* May 17, 1865, 2.

18. "Additional from the North," *New Orleans Tribune,* May 17, 1865, 2.

19. Adam H. Domby, "War within the States: Loyalty, Dissent, and Conflict in Southern Piedmont Communities, 1860–1876" (PhD diss., University of North Carolina at Chapel Hill, 2015), 331, 345–46.

20. For overviews of politics and Reconstruction, see Eric Foner, *Reconstruction: America's Unfinished Revolution, 1863–1877* (New York: Harper & Row, 1988); Du Bois, *Black Reconstruction in America;* Michael W. Fitzgerald, *Splendid Failure: Postwar Reconstruction in the American South* (Chicago: Ivan R. Dee, 2007); Mark Wahlgren Summers, *The Ordeal of the Reunion: A New History of Reconstruction* (Chapel Hill: University of North Carolina Press, 2014); Heather Cox Richardson, *The Death of Reconstruction: Race, Labor, and Politics in the Post–Civil War North, 1865–1901* (Cambridge, MA: Harvard University Press, 2004).

21. Eric Foner, *Freedom's Lawmakers: A Directory of Black Officeholders during Reconstruction,* rev. ed. (Baton Rouge: Louisiana State University Press, 1996).

22. "Negroes vs. White Men," *Daily Milwaukee News,* January 18, 1868, 4.

23. "Negro Cadets at West Point," *Intelligencer Journal,* February 10, 1868, 2.

24. "He Discriminates," *Idaho World,* July 22, 1869, 2; *Edgefield Advertiser,* July 7, 1869, 2.

25. For a short overview of many of the early West Point cadets, see William P. Vaughn, "West Point and the First Negro Cadet," *Military Affairs* 35.3 (1971): 100–102; and Wesley A. Brown, "Eleven Men of West Point," *Negro History Bulletin* 19.7 (April 1956): 147–57.

26. "Charleston," *New Orleans Tribune,* July 3, 1867, 3.

27. For more on Butler, see Elizabeth D. Leonard, *Benjamin Franklin Butler: A Noisy, Fearless Life* (Chapel Hill: University of North Carolina Press, 2022), esp. 182.

28. There is a reason that Lost Cause advocates worked so hard to erase Black military service in the decades after the war. It undermined their arguments against Reconstruction. For more on the attempts to erase Black military service, see Adam H. Domby, *The False Cause: Fraud, Fabrication, and White Supremacy in Confederate*

Memory (Charlottesville: University of Virginia Press, 2020), 31, 59, 124–25, 140–41, 145.

29. Pension file for Thomas C. Wilson (WC75600) Company K, 55th Massachusetts Colored Infantry, Case Files of Approved Pension Applications of Widows and Other Dependents of Civil War Veterans, ca. 1861–ca. 1910, Record Group 15, National Archives and Records Administration, Washington, DC, https://www.fold3.com/image/271458379.

30. "A Matter of Regret," *Detroit Free Press,* April 1, 1870, 2; Katharine Whittemore, "The Men Black Men of Amherst Left Out: Untold Histories of Black Alumni," *Amherst Magazine,* Spring 2021, https://www.amherst.edu/amherst-story/magazine/issues/2021-spring/bicentennial/the-men-black-men-of-amherst-left-out.

31. For an example of him listed as a former slave, see Patri O'Gan, "Duty, Honor, Country: Breaking Racial Barriers at West Point and Beyond," *National Museum of African American History and Culture,* May 2, 2022, https://nmaahc.si.edu/explore/stories/west-point. The article is otherwise excellent, but this mistake slips into numerous accounts of Smith.

32. Henry Ossian Flipper, *The Colored Cadet at West Point: Autobiography of Lieut. Henry Ossian Flipper, U.S.A. First Graduate of Color from the U.S. Military Academy* (New York: Homer Lee & Co., 1878), 314, https://docsouth.unc.edu/neh/flipper/flipper.html.

33. The 1860 census lists a carpenter named Israel Smith, married to a Catharine Smith (both listed as mulatto), with a ten-year-old son named James, who lived next to a member of the Guignard household (the family that reportedly enslaved Israel). In 1856, Israel was bequeathed to James Sanders Guignard Jr. on the death of his father. See Will of James Sanders Guignard, c1857, City of Columbia, Richland District, South Carolina, in "Index and Will, Vol. 3–4, Book L, 1787–1864," in South Carolina, Wills and Probate Records, 1670–1980, https://www.ancestry.com/discovery ui-content/view/664481:9080.

34. James Webster Smith, *Daily Phoenix,* May 11, 1867, 2; "Fair for the Benefit of the Republican Brass Band," *Daily Phoenix,* December 9, 1869, 3; "An Out and Outer," *Charleston Daily News,* March 23, 1870, 3.

35. E. P. Fyffe to E. M. Stanton, November 24, 1865, File H 1413-Hoge, Solomon Lafayette, Letters Received by the Commission Branch of the Adjutant General's Office, 1863–1870, National Archives, https://www.fold3.com/image/305065424; "Justices Willard and Hoge," *Yorkville Enquirer,* August 13, 1868, 1.

36. S. L. Hoge to Adjutant General, USA, August 3, 1868, File H 1413-Hoge, Solomon Lafayette, Letters Received by the Commission Branch of the Adjutant General's Office, 1863–1870, National Archives, https://www.fold3.com/image/305065480.

37. "The Situation in Columbia," *Charleston Daily Courier,* August 27, 1872, 2.

38. For more on Klan violence in South Carolina, see Bradley D. Proctor, "'The K. K. Alphabet': Secret Communication and Coordination of the Reconstruction-Era Ku Klux Klan in the Carolinas," *Journal of the Civil War Era* 8.3 (2018): 455–87; Elaine

Frantz Parsons, *Ku-Klux: The Birth of the Klan during Reconstruction* (Chapel Hill: University of North Carolina Press, 2015).

39. "South Carolina Outrages," *Harrisburg Telegraph,* October 24, 1868, 1.

40. "The Reign of Terror in the South," *Wheeling Intelligencer,* October 26, 1868, 3.

41. "Washington," *Detroit Free Press,* February 21, 1869, 1; House of Representatives, Papers in the Case of S. L. Hoge vs. J. P. Reed, Third Congressional District, South Carolina (41st Congress, 1st Session), House Misc. Doc. 18, 1869.

42. US Congress, *Testimony Taken by the Joint Select Committee to Inquire into the Condition of Affairs in the Late Insurrectionary States,* vol. South Carolina, vol. 2 (Washington, DC: Government Printing Office, 1872), 1256–60, https://quod.lib.umich .edu/m/moa/aca4911.0004.001/666; see also House of Representatives, Papers in the Case of S. L. Hoge vs. J. P. Reed, Third Congressional District, South Carolina (41st Congress, 1st Session), House Misc. Doc. 18, 1869.

43. "The First Negro Cadet," *Yorkville Enquirer,* June 9, 1870, 2.

44. "The Ball Opened," *Charleston Daily News,* July 4, 1870, 1.

45. "Congressman Hoge-Bottom Rail on Top," *Charleston Daily Courier,* August 10, 1870, 4.

46. Robert J. Schneller Jr. *Breaking the Color Barrier: The U.S. Naval Academy's First Black Midshipmen and the Struggle for Racial Equality,* (New York: NYU Press, 2007), 8.

47. "Expulsion of Whittemore," *Yorkville Enquirer,* March 3, 1870, 2. Additionally, both "The First Negro Cadet," *Yorkville Enquirer,* June 9, 1870, 2, and "State Items," *Yorkville Enquirer,* June 9, 1870, 2 appeared on the same day; one mentioned the corruption of Whittemore and the other concerned Black cadets.

48. Mr. Whittemore and His Expulsion," *Charleston Daily Courier,* February 25, 1870, 2.

49. "Butler, Roderick Random," History, Art & Archives: US House of Representatives, https://history.house.gov/People/Detail/10304; "Washington," *Chicago Tribune,* March 18, 1870, 1; "Congressional," *Detroit Free Press,* March 2, 1870, 4.

50. "The Capital," *Philadelphia Inquirer,* March 8, 1870, 1.

51. "Washington," *New York Herald,* March 1, 1870, 3; "Latest News by Mail," *Daily Picayune,* March 29, 1870, 9.

52. "Washington," *New York Herald,* March 1, 1870, 3.

53. "The Cadetship-Trader," *Valley Spirit,* March 2, 1870, 2; see also "Washington," *New York Herald,* March 1, 1870, 3.

54. "The Cadetship-Trader" *Valley Spirit,* March 2, 1870, 2.

55. "The Examination of Colored Cadets," *Georgia Weekly Telegraph and Journal and Messenger,* June 21, 1870, 8; "Local Items," *Daily Phoenix,* July 9, 1870, 2; "No. 15, Special Report in the case of Cadet James W. Smith," June 5, 1871, Reports on Examinations at the US Military Academy, RG 94, National Archives and Records Administration, https://www.familysearch.org/search/film/008812650?i=307&cat=354502.

56. William P. Vaughn, "West Point and the First Negro Cadet," *Military Affairs* 35.3 (1971): 100.

57. "The Colored Cadet from South Carolina," *Yorkville Enquirer,* July 14, 1870, 2.

58. The Examination of Colored Cadets" *Georgia Weekly Telegraph and Journal and Messenger,* June 21, 1870, 8.

59. "The Colored Cadets," *San Francisco Elevator* July 1, 1870, 2.

60. "How the Medical Board of West Point Got Rid of the Man and Brother," *Selma Dollar Times,* June 20, 1870, 2.

61. "Caste—West Point," *National Anti-Slavery Standard,* July 30, 1870, 1.

62. "The First Negro Cadet," *Yorkville Enquirer,* June 9, 1870, 2.

63. Andrew Johnson, "Prest. Johnson's Amnesty Proclamation," May 29, 1865, Library of Congress, https://www.loc.gov/resource/rbpe.23502500/?st=text.

64. "Amnesty," *Camden Journal,* January 5, 1871, 2.

65. "A Colored Cadet," *Republican Banner,* April 9, 1870, 4.

66. The Milk in the Cocoanut," *Selma Dollar Times,* April 11, 1870, 1.

67. For more on the concept of uplift suasion, see Ibram X. Kendi, *Stamped from the Beginning: The Definitive History of Racist Ideas in America* (New York: Nation Books, 2016).

68. "Color and Mormonism at West Point," *South Kansas Tribune,* June 21, 1871, 2.

69. "National Schools and Snobocracy," *New National Era,* August 6, 1874, 2.

70. Alan Friedlander and Richard A. Gerber, *Welcoming Ruin: The Civil Rights Act of 1875* (Boston: Brill), 248; see also "From Lagrange," *Atlanta Constitution,* September 12, 1872, 2; "The Georgia Press," *Georgia Weekly Telegraph and Journal and Messenger,* April 29, 1873, 8.

71. "The Georgia Press," *Georgia Weekly Telegraph and Journal and Messenger,* June 26, 1870, 8.

72. *Black Americans in Congress, 1870–2007* (Washington, DC: Government Printing Office, 2008), 88.

73. T. D. Allman, *Finding Florida: The True History of the Sunshine State* (New York: Atlantic Monthly Press, 2013), 260; Canter Brown Jr, *Ossian Bingley Hart, Florida's Loyalist Reconstruction Governor* (Baton Rouge: Louisiana State University Press, 1997), 207; "A Colored West Point Cadet Withdraws," *Daily Phoenix,* January 25, 1873, 3.

74. Peter D. Klingman, *Josiah Walls: Florida's Black Congressman of Reconstruction* (Gainesville: Library Press at University of Florida, 2017), 58–59.

75. "The Negro at West Point," *Daily Memphis Avalanche,* January 27, 1873, 1.

76. "Nicholas Muller, Old Politician, Dies," *New York Times,* Dec 13, 1917, 13.

77. "New York Notes," *Detroit Free Press,* August 26, 1877, 4.

78. "Race Prejudice," *Chicago Tribune,* January 26, 1878, 3.

79. Selected Documents Relating to Blacks Nominated for Appointment to the US Military Academy during the 19th Century, 1870–1887, NARA microfilm M1002, National Archives and Records Administration, https://www.familysearch.org/ark: /61903/3:1:3QHV-P3N2-599P-S?i=7&cat=354502.

80. Flipper, *Colored Cadet at West Point,* 265.

81. Eric Foner, *Reconstruction: America's Unfinished Revolution, 1863–1877* (New York: Harper & Row, 1988), 531; Ron Chernow, *Grant* (New York: Penguin, 2017). 768–70.

82. "Flipper," *Daily Democratic Statesman*, July 1, 1877, 2.

83. "The Negro Cadet Experiment," *Knoxville Daily Tribune*, June 26, 1877, 2.

84. "The Enactment of Equality—a Failure," *Morning Herald*, June 14, 1877, 2.

85. Donaldson, *Duty beyond the Battlefield*, 125.

86. *Pittsburgh Commercial*, October 21, 1876, 2.

87. Canter Brown, *Florida's Black Public Officials, 1867–1924* (Tuscaloosa: University of Alabama Press, 1998), 92–93.

88. Brown, *Florida's Black Public Officials*, 92–93; *Journal of the Proceedings of the Constitutional Convention of the State of Florida, Which Convened at the Capitol, at Tallahassee, on Tuesday, June 9, 1885* (Tallahassee: N. M. Bowen, state printer, 1885), 186, 316, 362, 381 562, 569.

89. For more on the Long Civil Rights Movement's framing of the past, see Jacquelyn Dowd Hall, "The Long Civil Rights Movement and the Political Uses of the Past," *Journal of American History* 91.4 (2005): 1233–63.

90. Scholars like Hilary Green place the end of Reconstruction later. Looking at education policy, she finds the major shifts came about later. See Hilary Green, *Educational Reconstruction: African American Schools in the Urban South, 1865–1890* (New York: Fordham University Press, 2016).

91. While never utilized, the Fourteenth Amendment's second section actually requires a state's congressional representation to be reduced if citizens are denied the vote "except for participation in rebellion, or other crime."

92. Paul E. Herron, *Framing the Solid South: The State Constitutional Conventions of Secession, Reconstruction, and Redemption, 1860–1902* (Lawrence: University Press of Kansas, 2017), 220.

93. *Black Americans in Congress, 1870–2007*.

94. "The First Negro Cadet," *Yorkville Enquirer*, June 9, 1870, 2.

95. "West Point," *Nashville Daily American*, April 7, 1880, 1; "Outrage at West Point," (Washington, DC) *People's Advocate*, April 10, 1880, 2.

96. Philip Dray, *Capitol Men: The Epic Story of Reconstruction through the Lives of the First Black Congressmen* (Boston: Houghton Mifflin Harcourt, 2008), 321–26.

97. *People's Advocate*, April 10, 1880, 2.

98. "South Carolina," *Weekly Louisianian*, May 1, 1880, 2.

99. Reprinted as "Action of the West Point Outrage," *Herald of Kansas*, April 30, 1880, 1.

100. "Hard on the Stalwarts," *Memphis Daily Appeal*, May 21, 1880, 2.

101. "Can Caste Be Conquered," *Buffalo Sunday Morning News*, June 4, 1882, 1.

102. "Lieutenant Flipper's Case," *The Inter Ocean*, February 13, 1882, 5.

103. "The Morning's Topics," *Buffalo Morning Express*, September 4, 1882, 1. The comment about scandal was likely also referring to Whittaker's failed cadetship.

104. "Cadet Alexander," *Omaha Daily World-Herald*, June 29, 1887, 1; "Cadet Alexander," *Daily Examiner*, June 14, 1887, 8.

105. *Macon Telegraph,* September 5, 1889, 4. For more on Young, see Donaldson, *Duty beyond the Battlefield,* esp. 126–49.

106. "Colored Georgian at West Point," *Savannah Tribune,* July 22, 1876, 1.

107. "Can Caste Be Conquered," *Buffalo Sunday Morning News,* June 4, 1882, 1.

108. "Senator Edmunds," *Springfield Patriot-Advertiser,* August 5, 1880, 1.

109. *Chicago Tribune,* September 6, 1880, 4.

110. "West Point," *Rutland Weekly Herald and Globe,* May 6, 1880, 2.

111. "Altogether Too Thin," *San Francisco Examiner,* May 1, 1880, 2.

112. "Hamburg Butler," *Inter Ocean,* April 27, 1880, 5.

113. "A Bystander's Notes," *Inter Ocean,* February 1, 1890, 4.

114. Wesley A. Brown, "Eleven Men of West Point," *Negro History Bulletin* 19.7 (April 1956): 147–57.

115. Blake Stilwell, "This African American 'Old Grad' Delivered a History of Race at West Point," Military.com, 2022, https://www.military.com/history/african -american-old-grad-delivered-history-of-race-west-point.html.

116. Aaron Morrison, Helen Wieffering, and Noreen Nasir, "'We Just Feel': Racism Plagues US Military Academies," AP News, December 2, 2021, https:// apnews.com/article/business-race-and-ethnicity-racial-injustice-army-only-on-ap -2975ab7e8d4fde2f275179e088878fb0.

117. Aaron Morrison, "Exclusive: These Alums Want West Point to Have an Honest Conversation about Race," *MIC,* May 18, 2016, https://www.mic.com/articles /143693/these-alums-want-west-point-to-have-an-honest-conversation-about-race.

118. David Shortell, "Black West Point Cadets under Scrutiny for Raised Fists in Photo," CNN.com, May 9, 2016, https://www.cnn.com/2016/05/08/us/west-point-cadets -photo/index.html; Sarah Sicard, "How Affirmative Action Works at West Point," *Task and Purpose,* July 19, 2016, https://taskandpurpose.com/news/race-factors -west-point-admissions/.

119. Corey Dickstein, "Recent West Point Grads Reveal Racist Incidents at Academy," *Stars and Stripes,* July 7, 2020, https://www.stripes.com/recent-west-point-grads -reveal-racist-incidents-at-academy-1.636694.

120. David Bindon, Simone Askew, Joy Schaeffer, Tony Smith, Cary Kehn, Jack Lowe, Netteange Monaus, Ashley Salgado, Maria Blom, "Policy Proposal: An Anti-Racist West Point," July 25, 2020, https://s3.amazonaws.com/static.militarytimes.com /assets/pdfs/1594132558.pdf.

121. Desiree D'Iorio, "Confederate Names Are Being Removed from Bases, but Is a KKK Image at West Point 'More Nuanced'?," Texas Public Radio, October 20, 2022, https://www.tpr.org/military-veterans-issues/2022-10-20/confederate-names-are -being-removed-from-bases-but-is-a-kkk-image-at-west-point-more-nuanced.

122. Irene Loewenson, "Military a Key Focus of Supreme Court Argument on Affirmative Action," *Air Force Times,* October 31, 2022, https://www.airforcetimes.com /news/your-military/2022/10/31/military-a-key-focus-of-supreme-court-argument

-on-affirmative-action/. For more on West Point admissions policies, see Sicard, "How Affirmative Action Works at West Point."

123. Philip Elliot, "Affirmative Action Still an Option at West Point, But Supreme Court Likely to Have Final Say," *Time,* July 6, 2023, https://time.com/6292620/affirmative -action-west-point-military-academies-supreme-court/.

124. Nina Totenberg, "Group Sues West Point, Seeking to Ban Affirmative Action in Admissions," NPR, September 19, 2023, https://www.npr.org/2023/09/19 /1200400504/west-point-affirmative-action; Anemona Hartocollis, "The Next Affirmative Action Battle May Be at West Point," *New York Times,* August 3, 2023, https://www.nytimes.com/2023/08/03/us/affirmative-action-military-academies .html.

125. Joshua Zeitz, "The Supreme Court Has Never Been Apolitical," *Politico,* April 3, 2022, https://www.politico.com/news/magazine/2022/04/03/the-supreme-court -has-never-been-apolitical-00022482; Rachel Shelden, "The Supreme Court Used to Be Openly Political. It Traded Partisanship for Power," *Washington Post,* September 25, 2020.

126. Meghna Chakrabarti and Tim Skoog, "Historian Heather Cox Richardson's Notes on the State of America," WBUR, September 29, 2023, https://www.wbur.org /onpoint/2023/09/29/historian-heather-cox-richardsons-notes-on-the-state-of -america.

127. Chakrabarti and Skoog, "Richardson's Notes on the State of America." For a basic introduction to the shift in the 1930s, see Richard G. Menaker, "FDR's Court-Packing Plan: A Study in Irony," Gilder Lehrman Institute of American History, https:// ap.gilderlehrman.org/history-by-era/new-deal/essays/fdr%E2%80%99s-court -packing-plan-study-irony.

128. Simon Clark, "How White Supremacy Returned to Mainstream Politics," Center for American Progress, July 1, 2020, https://www.americanprogress.org/article/white -supremacy-returned-mainstream-politics/.

129. For more on how racism plays a role in modern politics, see Josiah Ryan, "'This Was a Whitelash': Van Jones' Take on the Election Results," CNN.com, November 9, 2016, https://www.cnn.com/2016/11/09/politics/van-jones-results-disappoint ment-cnntv; Domenico Montanaro, "How the 'Replacement' Theory Went Mainstream on the Political Right," NPR, May 17, 2022, https://www.npr.org/2022/05 /17/1099223012/how-the-replacement-theory-went-mainstream-on-the-political -right.

MAKING WEST POINT "ALL-AMERICAN"

Integration as the Legacy of Civil War Service

Cameron D. McCoy

Civil rights activists Frederick Douglass and John Eaton Jr., a Union army chaplain, were keenly aware of the connection between citizenship, military service, and equitable formal education during the sectional crisis. Douglass and Eaton both believed that these three elements would be realized only through the crucible of war. According to Civil War activists, this realization of citizenship through military service would unlock the doors to greater social access, liberty, and education for newly freed Blacks and their posterity; the Black soldier had the most at stake in the Civil War. Douglass, a former slave, declared, "Once let the black man get upon his person the brass letters 'US,' let him get an eagle on his button and a musket on his shoulder and bullets in his pocket and there is no power on earth which can deny that he has earned the right to citizenship in the United States." During this same period, Eaton contended that "anyone devoted to his books was on the road to freedom; anyone ignorant of books was on his way back to slavery."[1] Both men provide telling assertions of the significance freed Blacks placed on military service and education during the postwar decades.

In addition to Douglass and Eaton, prominent leaders such as Henry Highland Garnet, William Wells Brown, Martin R. Delany, and George T. Downing recruited thousands of young Blacks for service in the Union army with the hope that their service would help transform the struggle into one that would liberate the enslaved and bring African Americans equal rights and greater social access in a transformed and redeemed republic.[2] For this reason, arguably, no other place in the United States during this period epitomized social access, liberty, and formal education for Black people like the United States Military Academy at West Point. If military service was the threshold to Black civic equality and liberty, as Douglass, Eaton, and others asserted, then a West Point education—and an army officer's commission—was an important marker on that path.

As the first and oldest US service academy, for many traditionalists, West Point is commonly viewed as *the* academy. Consequently, West Point holds special significance as a prestigious military institution in the evolution of American education. Thus, accessing West Point during Reconstruction became highly contested, especially for African Americans. Because West Point was first and foremost a military institution to prepare young officers for the US Army, Black wartime military service was seen as a viable path toward the promise of full citizenship.

The dilemma of educating Black Americans began with an admixture of questions regarding citizenship, myths, and dubious stereotypes about African Americans. In such an environment, the discourse about racial equality found particular meaning in military service. Therefore, the integration of West Point began with the American Civil War and the first Black servicemembers to don the Union Blue. In this sense, the legacy of Black Civil War service was dramatic in complex ways that went far beyond the battlefield.

The integration of West Point was influenced by the legacy of Black soldiers' Civil War service and how African Americans approached education as a way to uplift their communities and their race. Although this was a time of incredible uncertainty, white America understood and carefully guarded the social order of a country still committed in many ways to the antebellum racial status quo. Nonetheless, the war and its outcome opened possibilities for Black social progress despite white America's faith in sectional reconciliation. For Blacks and their moderate allies, educational coercion was a clear and recognizable path to realizing that promise. Black and integrated education during Reconstruction was framed and informed by Black experiences of the Civil War and its aftermath. The United States Colored Troops (USCT) and integration at West Point were thus thoroughly intertwined, with the latter being the legacy of the former. The relationship between Black wartime service and later attempts to integrate West Point is an important element in understanding postwar integration as a lost opportunity, one that went beyond the success or failure of individual cadets.

Before, during, and after the Civil War, many white Americans saw the specter of an educated and armed Black population as a haunting image. Yet African Americans placed their faith in education as the pathway to true liberty and more complete access to citizenship. However, in practice, this freedom (and the related emphasis on education) was contested, particularly when it came to Blacks accessing the rudimentary elements of life entitled to citizens of a democratic republic (e.g., the right to serve in the military, the freedom to move about, the right to choose their employer, and freedom from physical punishment and the breakup of families). Hilary Green, William H. Watkins, and Heather Andrea Williams all provide accounts that place the story of Black education during Reconstruction at the base of political and ideological foundations rooted in the power of literacy. According to these authors, the power of literacy spearheaded claims that education was a civil right and demonstrated its transformative power among both Black and white populations; moreover, education punctuated Black Americans' claims to social equality.[3]

In freedom, Blacks moved to protect their families and gain educational rights. Even so, centuries of bondage and slavery had a broad and lasting impact, and many Americans harbored something between a sense of apathy and profound discomfort about civic and social harmony between the races. More importantly, restrictive laws (i.e., Black Codes) had sealed off even the most common spaces for basic educational access for freed Blacks after the Civil War. This is why some northern educators—dedicated to the moral and social regeneration of freed Blacks—understood the power of a Union victory and positioned themselves to be the first to usher in a world of integrated education. Inspired by abolition, some sought to further dissolve barriers to military service, to institute formal education, and to open greater avenues to universal liberty in kind, as well as full-fledged citizenship under the Constitution.[4]

For many newly freed Blacks, military service was the most realistic access point to citizenship—one that would position them to claim general social equality with white Americans. Military service in America has traditionally held special meanings for authentic claims to citizenship. For many nineteenth-century Americans, military service represented sacrifice, unselfishness, valor, and the opportunity to demonstrate bravery and heroism in the face of grave danger on behalf of the nation. Black Americans saw this opportunity—previously denied to them—as a way not only to defeat those who would enslave them but also to assert and secure claims to full citizenship.

The USCT demonstrated how the Civil War had broken and then remade the United States in profound ways. In 1861, northern whites voiced widespread opposition to the use of Blacks as soldiers, but the hard experience of war soon dulled this opposition. Illinois, for example, which had only recently banned Blacks from even entering the state, slowly lifted racial restrictions following the admirable performance of the USCT in several well-publicized engagements. While several states also moved sluggishly when it came to *equal protection under the law,* in Illinois civic responsibility followed military service as a natural progression. As a result, Blacks were allowed to serve on juries and provide courtroom testimony—in no small part due to USCT achievements on and off the battlefield. Among many whites, as the war ended, the matter of egalitarianism as a characteristic of citizenship varied from state to state. But there were limits to the degree of egalitarianism that USCT battlefield exploits could inspire in white Americans, even in the North. As historian Eric Foner has pointed out, even as most northern Republicans came to consider "equality in civil rights—equal treatment by the courts and civil and criminal laws" to be "nearly essential" in order to secure "an individual's natural rights," most of them still would not concede full social equality to Black Americans.[5] For most white Americans, there was a tension between recognizing Black claims to rights—whether civil, natural, political, or social—and actually granting them.

Some of the first regiments of free Blacks came into service in New Orleans in September 1862 through the efforts of Gen. Benjamin F. Butler. General Butler's three regiments of African Americans (Louisiana Native Guards) were among the first to formally muster into Union ranks. North Carolina, South Carolina, and Kansas also organized early Black units that proved to be successful in raids and skirmishes. Early in the war Lincoln had opposed the idea of Blacks fighting for the Union, but he later altered his view. Lincoln declared that slaves in states still in rebellion on January 1, 1863, "shall be then, thenceforward, and forever free," and further proclaimed that freed Blacks "would be received into the armed service of the United States."[6] Against the terrible backdrop of the war's mounting losses, Lincoln planned to tap into a new source of fighting men, "the great available and yet unveiled of, force for the restoration of the Union," tacitly connecting the previously distinct issues of freedom and military service.[7]

On a strategic level, Lincoln sincerely believed this would both weaken the South and strengthen the Union. Black recruitment took laborers from the South and enrolled them in the Union army in places that otherwise would have been filled by white men. But heroic battlefield performance by Black regiments in the spring and summer of 1863 influenced another shift in Lincoln's thinking. He came to see Black soldiers as more than just a new pool for military labor; rather, they were soldiers in every sense of the word, who deserved the full faith and protection of their government. A week after Col. Robert Gould Shaw led the all-Black 54th Massachusetts Infantry Regiment into battle at Fort Wagner, "[Lincoln] sent for General Halleck and told him to prepare an order for the protection of Negro prisoners. The draft submitted by the War Department reflected Lincoln's views, ordering that, for every

Union soldier killed in violation of the laws of war, a rebel soldier would be put to death, and that, for every Union soldier sold into slavery, a rebel soldier would be put at hard labor on the public works."[8] Confederate commanders were incensed when it came to recognizing Blacks as anything other than property. "The Richmond government could not bring itself to return to the Union armies a Negro who was some Southerner's property. Moreover, to exchange a black man for a white man was to be a party to racial equality."[9]

With views of rights and obligations in such flux, staunch abolitionists began recruiting freed Blacks with haste. But once recruitment started, the numbers quickly expanded. What began in 1862 as localized recruiting efforts on the part of a small number of Union department commanders who were acting beyond War Department instructions had by the end of 1863 turned into a formidable body of sixty regiments fully mustered into federal service. And more would follow in 1864 and 1865.[10]

Even with such growth in Union manpower, African Americans were still considered by many civilian and military observers to be second-class soldiers. As Union armies broke through Confederate forces in the South, thousands of enslaved Blacks fled their enslavement, an event that much of white America struggled to process. The image of freed Black slaves challenged white racial sensibilities amid the stark social changes that would accompany granting and acknowledging equitable citizenship to a Black populace.[11] Many whites—even in the North—were discomfited by the sight of armed Black men. But like Lincoln, some changed their views over time—especially in light of such atrocities as the massacre at Fort Pillow, in which Confederate troops refused quarter for surrendering USCT soldiers and murdered scores in cold blood. This slaughter was indeed needless under the veneer of war, as "many Negroes were put to death while holding their hands above their heads in a token of surrender. The losses among colored troops were exceptionally heavy." Confederate general Nathan Bedford Forrest would later boast that "the river along the fort was dyed with blood for 200 yards." Horrified by Forrest's actions, the North saw the carnage at Fort Pillow as "racial antagonism at its worst."[12] Outrages like this committed by southerners made it difficult for Black soldiers to recognize if they would be captured in battle as prisoners of war or as slaves in insurrection—a very real and fearsome question that was not an issue for their white counterparts.

Due to USCT service and sacrifice, the image of Black Americans as fellow human beings began to shift for at least some northern whites, even if begrudgingly. Many white northern soldiers had primarily grown up knowing and understanding the Black man as a distorted caricature who had been portrayed as disabled and irredeemable. In contrast, what they found was a real human, struggling to be in control of his current and future destiny. "The black folks are awful good, poor miserable things that they are. The boys talk to them fearful and treat them [whatever way they want] and yet they can't talk two minutes, but tears come to their eyes, and they throw their arms up and praise de Lord for de coming of de Lincoln soldiers," as described by one Wisconsin soldier after fighting in battle with Black soldiers. Attitudes revised during the shared endurance of war reflected a marked departure for many white soldiers.[13]

Officers' views proved similarly malleable. In July 1862, Charles Francis Adams Jr. wrote emphatically: "The idea of arming the blacks *as soldiers* must be abandoned. . . . The Negro is wholly unfit for cavalry service."[14] Yet one year later, during the decisive month of July 1863, he confidently wrote to his father: "The Negro regiment question is our greatest victory of the war so far, and, I can assure you, that in the army, these [men] are so much a success that they will soon be the fashion." What shifted Colonel Adams's perspectives during those twelve months from formal castigation to lauding approval? It was a changing perspective that could only have been prompted by firsthand observation, an encounter with the reality that Black men could perform as bona fide soldiers when given the opportunity. By the summer of 1864, as a lieutenant colonel, Adams had taken command of the Fifth Massachusetts (Colored) Volunteer Cavalry and then led Black troops into battle during the Siege of Petersburg, Virginia. Clearly, for him, Blacks had demonstrated their value as soldiers and earned some element of respect as men also.[15]

In the spring and summer of 1863, Black regiments fought valiantly in several notable battles. The recognition of the value of the Black man as a soldier of the Union did not come from the final clarification of administration policy or the activities of advocates and recruiters. Of course, these were important; without them the entire movement to arm Black men may have died before it took form. No matter the level of encouragement or endorsement the administration and the Department of War applied to the Black soldier policy, the complete acceptance of Black men as soldiers ultimately depended on their performance in battle.[16]

The first major engagement during the Civil War in which Black soldiers participated was the Battle of Port Hudson on May 27, 1863. Control of the Mississippi River was a primary federal objective, and by the spring of 1863 the Union army had secured the entire course of the river except for Vicksburg, Mississippi, and Port Hudson, Louisiana—both Confederate bastions. To cut off the Confederates in the Trans-Mississippi West and provide midwestern farmers with a water route for crop shipments, the Department of War determined that units under the command of Maj. Gen. Ulysses S. Grant were to capture Vicksburg; and Maj. Gen. Nathaniel P. Banks, commanding the Department of the Gulf, was to seize Port Hudson. If successful, the campaign would effectively split the Confederacy in two.[17]

Despite Banks's terrible defeat in his attempt to take the Confederate stronghold, the Battle of Port Hudson marked a turning point in attitudes concerning the use of Black soldiers. Maj. Gen. Henry W. Halleck stated, "The severe test to which they were subjected, and the determined manner in which they encountered the enemy, leaves upon my mind no doubt of their ultimate success." Brig. Gen. Daniel Ullman concurred enthusiastically. Ullman, who was in the process of raising an African American brigade in Louisiana, informed Secretary of War Edwin Stanton, "The brilliant conduct of the colored regiments at Port Hudson, on the 27th has silenced cavilers and changed sneers into eulogizers. There is no question but that they behaved with dauntless courage."[18]

The *New York Times,* which had cautiously endorsed the limited use of Black troops on a trial basis just four months before the assault on Port Hudson, now declared the experiment a resounding success: "Those black soldiers had never before been in any severe engagement. They were comparatively raw troops, and subjected to the most awful ordeal that even veterans ever have to experience—the charging upon fortifications through the crash of belching batteries. The men, white or black, who will not flinch from that will not flinch from nothing. It is no longer possible to doubt the bravery and steadiness of the colored race, when rightly led."[19] Black troops won respect with their battlefield performance, even among those who previously doubted their innate ability to serve as soldiers alongside whites.

The success of Black soldiers at Port Hudson and the insatiable need of the Union army for fresh soldiers encouraged the large-scale enlistment of Black men. These soldiers also discovered a newfound social power in the execution of their military duties. When African American soldiers endured their baptism in blood in the late spring of 1863, their contributions in battle disproved the racist ideas that Black men were cowardly by nature and lacked either the discipline or intelligence to succeed in combat. While this did not produce immediate converts and believers among the many, it still *disproved* what many considered the science behind Black performance, bravery, and valor. Such ingrained cultural notions in a racially defined country died hard. Because of racial discrimination and prejudice against them, the use of Black soldiers in battle was largely limited to units from states that pressed for them to be used in combat or in places where military commanders were willing to employ them or could not dispense with their services. The employment of Black troops in battle was no social experiment—it was an expedient nod to the war's insatiable demand for men by the many thousands.[20]

Two exponents of Black manpower, Union generals Lorenzo Thomas (a West Point graduate) and Daniel Ullman (commanding general of the Corps D'Afrique) would have labored in vain if the regiments they raised failed to measure up as legitimate soldiers in battle. After the pronouncement of the Emancipation Proclamation, the governor of Massachusetts, John Albion Andrew, received permission to raise a Black regiment. Andrew's higher purpose could never have been realized if the 54th Massachusetts Infantry had faltered before the earthworks of Fort Wagner in July 1863. Black troops not only faced an enemy dedicated to the belief that the proper place for them was in slavery, but they also confronted doubts about their fighting abilities among white northerners—including their counterparts in Union armies.[21]

Given the conservative nature of many long-serving Union officers, several could not be swayed to accept Black soldiers; many remained inclined to use them only as auxiliaries, laborers, or guards rather than as combat troops. No amount of talk could win the African American soldiers' place in the Union army better than their performance under fire—the truest test of a soldier's mettle. The Black soldier had to prove himself (in soldierly fashion) by fighting courageously and, if necessary, dying in battle. From 1863–65 that is precisely what Black troops did, serving with distinction at Port Hudson, Milliken's Bend, Vicksburg, Fort Wagner, and Petersburg, Virginia, among other engagements.[22]

Since the 1600s, the practice of slavery had warped the hearts and minds of the American people. Mid-nineteenth-century Americans had grown up in awe of slavery and with racial contempt for Blacks, declaring that the country would not have Black soldiers. *Harper's Weekly* reported, "It is very cheering to believers in human progress, and to men who honestly admit that the world moves, to perceive that the short period of two years has sufficed to cure an evil of so long-standing, and has educated even the most hunkered Democrat of 1861 into a willingness to arm the blacks."[23] The battle of Port Hudson was the first event that marked a noticeable shift in attitudes concerning Blacks in uniform and Black troops' performance. This engagement was also a significant element in bringing about changes in broader perspectives connected to the service and valor of Black soldiers.

Despite an overwhelming defeat at Port Hudson, the grit demonstrated by Black veterans of the battle had to be respected. In the face of terrible odds, Black troops distinguished themselves, while whites grudgingly took note. Although prejudice remained, after observing the valiant performance of African American soldiers in the First and Third Louisiana Native Guards, a white New Yorker commented: "They charged and re-charged and didn't know what retreat meant. They lost in their two regiments some four hundred as near as

I can learn. This settles the question about niggers not fighting well. They, on the contrary, make splendid soldiers and are as good fighting men as any we have."[24] Writing with considerable prescience, a Massachusetts soldier agreed with this assessment, asserting, "A race of serfs stepped up to the respect of the world [at Port Hudson] and commenced a national existence."[25]

While these remarks indicate that eradicating bigoted attitudes had a long road ahead, the improving reputation of Black soldiers ran through the ranks and up the chain of command. Writing to his wife, Union colonel Benjamin H. Grierson remarked, "The Negro regiments fought bravely yesterday.... There can be no question about the good fighting qualities of Negroes hereafter."[26] African American soldiers proved themselves in battle, and their actions spoke for them on all levels of Union commands. On Tuesday, June 9, 1863, in an official report from Union general Nathaniel Banks regarding the effectiveness of the Black troops in the siege of Port Hudson, the *New York Times* reported:

> In speaking of the Negro troops, [General Banks] says: "They answered every expectation. Their conduct was heroic. No troops could be more determined or more daring. They made, during the day, three charges upon the batteries of the enemy, suffering very heavy losses, and holding their position at nightfall with the other troops.... The highest commendation is bestowed upon them by all the officers in command.... [Whatever] doubt may have existed before as to the efficiency of organizations of this character, the history of this day proves conclusive to those who were in a condition to observe the conduct of these regiments, that the [Federal] Government will find in this case of troops effective supporters and defenders."[27]

General Banks's reference to the attack on Port Hudson was similar to previously published reports. Although Union forces lost approximately one thousand troops during this battle, Black soldiers engaged the enemy courageously, leaving no doubt in the minds of skeptics that they too were equals. And if they could be equals in battle, this meant—for some, not all—they could be equals in all areas of life, including higher education if allowed access. However, becoming an excellent soldier required good officers and commands with careful discipline to properly guide these men within the ranks. This part of the narrative would be the most controversial when it came to granting Blacks access to the rights of formal education; in this area, as in others, Blacks would always have to exercise caution because dangers such as lynching were only a glance, a gesture, or a word away from execution.

Unlike the *New York Times,* which before General Banks's comments had only cautiously endorsed the limited use of Black troops, *Harper's Weekly* sanctioned the enlistment and full participation of Black troops during the war, commending their performance at Port Hudson: "The magnificent behavior of Negro troops at Port Hudson recalls the fact that it is just two years since a warning, stated in the columns of this paper, 'that if this war lasted we should arm the Negroes, and use them to fight the rebels,' was received with shrieks of indignation, not only at the South and in such semi-neutral States as Maryland and Kentucky but throughout the loyal North and even in the heart of New England." The article further observed, "At that time, the bulk of the people of the United States entertained a notion that it was unworthy of a civilized or a Christian nation to use in war soldiers whose skin was not white."[28] Notwithstanding, Blacks remained driven to accept the burdens attendant with obtaining equal shares of full-fledged citizenship.

That African American soldiers fought bravely at Port Hudson and in the famous assault at Fort Wagner speaks to recognizing that even when they fought in a failing effort, they demonstrated fortitude and could win a moral victory. This courage and determination won them the admiration of white officers and soldiers, and of *some* members of the northern public who read about their exploits in the newspaper. By the end of the war, the army recognized their valor by awarding Black soldiers many decorations, including sixteen Congressional Medals of Honor. Nevertheless, their wartime performance failed to bring the lasting social change desired by those who sacrificed. Eric Foner, for example, highlights how after the war the Fourteenth Amendment "ostensibly affected" both races "equally" when it came to dealing with criminal offenses; yet according to a Tennessee Black convention, punishments among whites and Blacks were treated with extreme bias under the law.[29] But even if the nation's pervasive racial bigotry was not swept aside at war's end, the role of the Black soldier still influenced moderate Republicans to believe that the federal government should guarantee the equality of all citizens before the law in order to ensure their independence and self-sufficiency. After the war, politicians and federal officials took small but significant steps toward easing the color line that hindered antebellum Black education and liberty.

In 1866, congressional legislation opened a new chapter in American history and allowed Black Americans to play a major role in the development of the West, which first began with their contributions and struggles before and during the Civil War. The 39th Congress passed an act to increase and fix the Military Peace Establishment of the United States, and six all-Black regiments became part of the US military. Historians and scholars agree that the role of Black fighting men in the Civil War was important to northern efforts and postwar developments. They also agree that the most important result of their wartime contribution was the advancement of Black claims on social equality. This idea of freedom and social equality gave great confidence and pride to these long-oppressed people.[30] However, in 1865 it remained unclear what the "elusive term 'equality' [meant]" for many northerners. Massachusetts senator Henry Wilson declared, "I believe in equality among citizens—equality in the broadest and most comprehensive democratic sense," and yet this work was still unfulfilled at the time, and it would demand many more decades of hard-fought battles in the name of true social progress.[31]

Earlier in the war the Union army had nearly defined "equality" after capturing the main harbor of the Sea Islands off the coast of South Carolina. The Port Royal Experiment stands out as a testament to Black independence and self-governance. Liberated by Union troops in 1861, Port Royal Harbor and the Sea Islands served as a joint governmental and private effort initiated during the early stages of the war to transition freedpeople to liberty and self-sufficiency. After white residents fled the area, several private northern charities helped roughly ten thousand freed Blacks to become self-sufficient—a model of how the aims of Reconstruction might have taken shape for the nation. At Port Royal African Americans demonstrated the ability to live independently, free of white control, and prosper when given the ability to self-govern. Although the Port Royal Experiment eventually came undone due to white reconciliation after the war, the model served as a blueprint for independence free from gratuitous bias and exclusion.[32]

Exploring the relationships and social education that took place between Black and white soldiers—officers and enlisted—from 1861 until the early period of Reconstruction is important to better understanding integration at West Point during Reconstruction.[33] The army that West Point presented for Black male applicants would indeed be segregated.[34] Yet Henry Ossian Flipper (the first Black graduate of West Point) took pride in his pluckiness and what that meant in connection to being admired for his resolve vis-à-vis social equality.[35] It was clear to many African Americans that they were risking their lives to win not only their very freedom but also to establish themselves as contributing members of a society fixated on marginalizing their existence as full-fledged citizens.

In the Civil War's wake, the so-called Radical Republicans and their political allies labored for the admission of Blacks into all public spaces of American life. Of course, there was considerable opposition to breaking the status quo in the country's racial and social hierarchy. Most white Americans—regardless of political affiliation or even sectional identity—struggled mightily to genuinely accept the notion that Blacks were their equals. After the war, there were no official laws specifically excluding African Americans from applying to or attending the institution, which made West Point stand out among other universities at the time. However, "since these appointments were to be made by members of Congress, there was a political price involved: no member insecure in his job would have appointed a black man from a district whose voting population was white, and among whom blacks were often looked down upon and even despised." The appointment of Black men to West Point's corps of cadets was not an action taken lightly; it was a deliberate and potentially costly political gambit on the national stage, as Adam Domby has pointed out in the first essay in this volume.[36]

Admission into West Point carried with it, at least for members of the newly freed Black community, strict criteria that many, if not most, could not meet. If selected, incoming cadets were first advised in writing—this simple measure of being able to read eliminated a large swath of potential Black candidates—of the age and height requirements for admittance. All incoming cadets (Black and white) also had to be "free of any infectious or moral disorder, and, generally, free from any deformity, disease, or infirmity which may render them unfit for arduous military service."[37] For a Black cadet to have been *free of* the subjectivity this list would present to any unbiased medical professional would be one mean feat, especially when defining a Black man's capability to be fit for "arduous military service" (a nebulous term)—a complication not faced by many, if any, of their white counterparts.

Henry O. Flipper had been formerly enslaved, so for him to be "proficient in Reading and Writing; in the elements of English Grammar; [and] in Descriptive Geography" separated him from the majority of the newly freed Black population. Most Blacks did not possess this menu of academic knowledge.[38] There were several more markers West Point outlined as "disqualifiers" (e.g., feeble constitution and muscular tenuity; unsound health from whatever cause; loss of many teeth, or the teeth generally unsound; impediment of speech; flat feet, bunions). These conditions also were problematic even for the healthiest and most sound-minded middle-class young white man.[39] For a former slave, such challenges could be nearly insurmountable during Reconstruction.

Understanding the highly political act of nominating a Black man to West Point, former Union general Benjamin F. Butler, who now represented the Fifth Congressional District of Massachusetts, became the first congressman to consider appointing Blacks to West Point. Butler, like some of his contemporaries who fought in the war, believed this would be a tremendous accomplishment in helping to break down old views of Black performance and skill in the arena of higher education.[40] In 1870 General Butler's efforts allowed for Charles Sumner

Wilson (Massachusetts), Michael Howard (Mississippi), and James Webster Smith (South Carolina) to receive nominations to West Point.[41]

Michael Howard initially caught the attention of the larger public. An article in the *New York Sun* entitled "The West Point Revels" showed that white America was not yet ready to see this level of social progress by the Black population. West Point was still considered a sacred space and a bastion reserved for young white men. Howard's arrival at West Point was such a departure from the social status quo that it caused a sort of cognitive dissonance among whites. "There had been rumors that negro boys had been appointed to the National Academy, but the absolute arrival of an African, commission in hand, is too much for West Point human nature to endure."[42] Despite the progressive attempts of many abolitionists and visionaries who wanted a world that centered equality at the highest levels, "aristocratic professors and jaunty cadets [were] speechless. The time for breaking forth of their indignation [had] not yet arrived. They [could not] not do the subject justice, but their indignant countenances and ominous looks [indicated] the coming storm."[43]

This storm of radical change would be nothing more than a mild gust of wind, but white cadets believed that the sky was falling! Many were ready to physically assault, and even kill, Howard. Others even threatened to resign. Although this captured the nation's imagination, Howard's fame would be short-lived. He would not enter West Point due to multiple failures on entrance examinations given shortly after cadets' arrival for plebe summer. Howard was unable to pass course exams in subjects like arithmetic, reading, writing, grammar, geography, and history, so he never matriculated with other nominees to the new class.[44]

James Webster Smith had minimal problems with the preliminary exams and so became the first Black cadet at West Point, though he eventually would be dismissed—not a fate dissimilar to that of many white students.[45] Despite Butler's vision, in a manner that reflected the initial experiences of their Civil War predecessors, the societal shift would be slow for white America. Blacks were still viewed as second-class citizens and struggled to escape the narrative of their ineptitude in every professional space freely available to white Americans. Howard's departure left Smith estranged from his white counterparts and, at least in spirit, entirely alone until Henry Ossian Flipper arrived in 1873.

Flipper's arrival at West Point marked a turning point in many ways. First, he was eager to attend West Point; Flipper understood that the academy was the summit he needed to surmount to enjoy life's most treasured rewards. Second, he had been limited in life and knew that West Point would provide him with unparalleled opportunities, allowing him to enter a hallowed space and fraternity that would remain solely his for life. "May 20th, 1873! Auspicious day! From the dock of the little ferry-boat that steamed its way across from Garrison's on that eventful afternoon I viewed the hills about West Point, her stone structures perched thereon, thus rising still higher, as if providing access to the very pinnacle of fame, and shuddered," declared Flipper. This first view expressed a young man's eagerness to embark on a dangerous yet fulfilling journey. Flipper concluded, "With my mind full of the horrors of the treatment of all former cadets of color, and the dread of inevitable ostracism, I approach tremblingly yet confidently."[46]

West Point left James Webster Smith bitter after his dismissal. According to Flipper, Smith endured incredible pain and suffering for four years, emanating from white cadets' responses to his presence, which were motivated by and emblematic of the racist attitudes held by the majority of white Americans at the time. Until the close of World War II, white cadets would talk to Black cadets only in the course of "official business," a practice that became standard treatment for all African Americans at West Point. The horrendous experiences of pioneering Black cadets failed to truly shift white bias and disapproving stereotypes, even if their successes challenged racial misconceptions. The status quo might have been breached but would not go peaceably into the past.

Maj. Gen. Hanson Edward Ely and Brig. Gen. Lytle Brown's evaluation of the 1925 Army War College study, *The Use of Negro Manpower in War,* damned African Americans in the military until the end of the war in Vietnam. The report canonized every negative generalization one could assign to a cowardly group during combat, emphasizing the limitations of African Americans' abilities to occupy significant military leadership positions, unjustly focusing its findings on Blacks' dependence on the white race to achieve military success, which ultimately emboldened white assumptions that Black men would continue to make poor soldiers, carrying with it twisted conclusions about their military discipline and conduct. The 1925 study emerged as one of the guiding documents for analyzing and sustaining "the physical, mental, moral and psychological qualities and characteristics of the negro as a sub-species of the human family."[47]

The "findings" of these types of studies stemmed primarily from the Civil War era's overt bigotry, and for Black cadets, simply *existing while black* proved to be an extreme sport. The burden of Black skin in white America was enough to prompt relentless distress, and questions such as "Am I worthy to be here?" and "Will this White cadet kill me if I make eye contact?" were forever relevant, or at least they seemed to be so to young Black cadets amid their white peers. Even so, Flipper remained steadfast and committed to the higher laws of humanity and West Point's many tenets. Even after enduring nearly complete ostracism, Flipper's remembrance of his West Point years reflects a young man who recognized the immense opportunity that accompanied his accomplishment in the context of a racially bound society:

> I have not a word to say against any of the professors or instructors who were at West Point during the period of my cadetship. I have everything to say in their praise, and many things to be thankful for. I have felt perfectly free to go to any officer for assistance, whenever I have wanted it, because their conduct toward me made me feel that I would not be sent away without having received whatever help I may have wanted. All I could say of the professors and officers at the Academy would be unqualifiedly in their favor.[48]

Certainly, the fate of the formerly enslaved was tied up with Union victory, emancipation, and the Thirteenth Amendment; these men understood that they were fighting for the freedom of their race, for equality, and for civil rights.[49] After the war, John Eaton's assertion that "anyone devoted to his books was on the road to freedom; anyone ignorant of books was on his way back to slavery" proved a telling assertion of the significance assumed by education for freed Blacks.[50] In evaluating the integration of West Point within a broad context of Black and integrated education, Flipper—on the route blazed by James Webster Smith, even in failure—took Eaton's words to heart. In the face of insuperable hostility and uncertainty, Flipper maintained a devotion to his studies that blocked out the noise of bigotry and ignorance, clearing a path directly to self-determination through West Point in 1877.

As of 1898, "the War Department [has] kept no record of the colored cadets separate from the white; nor do the nominations, or appointments, of cadets make any distinction in the respect. The Department [is] therefore unable to inform you of the actual number, etc., of colored youths appointed as cadets; but it [is] believed that 12 colored cadets have been admitted to the U.S. Military Academy," replied the War Department's adjutant general when asked by Titus N. Alexander of the *Detroit Tribune* for the names of Black cadets who had received appointments to West Point.[51] Because the War Department did not maintain a separate repository for Black men who attended West Point, it is difficult to surmise the actual number that donned the cadet gray in any formal way during the nineteenth century.[52] But regardless of the number or the army's flimsy attempt to appear color-blind in its recordkeeping, educating young Black men brought into sharp focus the legacy of Black soldiers who shed negative stereotypes and claimed respect for their service and sacrifice on Civil War battlefields. The accession of Black cadets at West Point during the late nineteenth century was itself an expression of wartime claims of martial equality, a forceful and resounding expression that shook white America awake to the basic humanity of Black Americans.

Notes

1. Frederick Douglass quoted in William A. Gladstone, *United States Colored Troops, 1863–1867* (Gettysburg, PA: Thomas Publications, 1990), 13, 118–21. See also Dudley Taylor Cornish, *The Sable Arm: Black Troops in the Union Army, 1861–1865* (Lawrence: University Press of Kansas, 1987); Joseph T. Glatthaar, *Forged in Battle: The Civil War Alliance of Black Soldiers and White Officers* (New York: Free Press, 1990), 136, 250; US Commissioner of Education John Eaton quoted in John Eaton, *Grant, Lincoln, and the Freedmen: Reminiscences of the Civil War* ([1907]; Bellevue, WA: Big Byte Books, 2016), 208; Paul David Phillips, "Education of Blacks in Tennessee during Reconstruction, 1865–1870," *Tennessee Historical Quarterly* 46 (Summer 1987): 98. Eaton received a commission as colonel of the 63rd United States Colored Infantry (USCI) Regiment, serving with the regiment until being appointed assistant superintendent of a district for the Freedman's Bureau, which he held until his resignation in 1865. In the postwar years, Eaton dedicated himself to the establishment and advancement of a public education system; see Brian Kilmeade, *The President and the Freedom Fighter: Abraham Lincoln, Frederick Douglass, and Their Battle to Save America's Soul* (New York: Sentinel, 2021), 199–201.

2. Joyce Hansen, *Between Two Fires: Black Soldiers in the Civil War* (New York: F. Watts, 1993), 67–78.

3. See Hilary Green, *Educational Reconstruction: African American Schools in the Urban South, 1865–1890* (New York: Fordham University Press, 2016); William H. Watkins, *The White Architects of Black Education: Ideology and Power in America, 1865–1954* (New York: Teachers College Press, 2001); Heather Andrea Williams, *Self-Taught: African American Education in Slavery and Freedom* (Chapel Hill: University of North Carolina Press, 2007).

4. Phillips, "Education of Blacks in Tennessee," 98. See also Green, *Educational Reconstruction;* Watkins, *White Architects of Black Education;* and Williams, *Self-Taught.*

5. Eric Foner, *Reconstruction: America's Unfinished Revolution, 1863–1877* (New York: HarperCollins, 2005), 231; Phillips, "Education of Blacks in Tennessee," 98.

6. Emancipation Proclamation, January 1, 1863, Record Group 11: General Records of the United States Government, 1778–2006, Series: Presidential Proclamations, 1791–2016, National Archives and Records Administration (NARA), Washington, DC (hereafter cited as Emancipation Proclamation).

7. Lincoln quoted in Howard C. Westwood and John Y. Simon, *Black Troops, White Commanders, and Freedmen during the Civil War* (Carbondale: Southern Illinois University Press, 1992), 37–54. See also James M. McPherson, *The Negro's Civil War: How American Negroes Felt and Acted during the War for the Union* (New York: Vintage Books, 1965), 37–68; John David Smith, ed., *Black Soldiers in Blue: African*

American Troops in the Civil War Era (Chapel Hill: University of North Carolina Press, 2004), xiii, xiv, xvii, 22–23.

8. Benjamin Quarles, *Lincoln and the Negro* (New York: Da Capo Press, 1990), 174–75, 179–83. For more on Union colonel Robert Gould Shaw, see Shaw, *Blue-Eyed Child of Fortune: The Civil War Letters of Colonel Robert Gould Shaw,* ed. Russell Duncan (Athens: University of Georgia Press, 1992); Russell Duncan, *Where Death and Glory Meet: Colonel Robert Gould Shaw and the 54th Massachusetts Infantry* (Athens: University of Georgia Press, 1999).

9. Quarles, *Lincoln and the Negro,* 175.

10. Westwood and Simon, *Black Troops, White Commanders, and Freedmen,* 37–54; McPherson, *Negro's Civil War,* 37–68; Smith, *Black Soldiers in Blue,* xiii, xiv, xvii, 22–23. Regarding literacy among freed Blacks, estimates suggest that between 5 and 10 percent of freed Blacks could read. See James D. Anderson, *Education of Blacks in the South, 1860–1935* (Chapel Hill: University of North Carolina Press, 1988); Ronald E. Butchart, *Schooling the Freed People: Teaching, Learning, and the Struggle for Black Freedom, 1861–1876* (Chapel Hill: University of North Carolina Press, 2010); Paul A. Cimbala and Randall M. Miller, eds., *The Freedmen's Bureau and Reconstruction: Reconsiderations* (New York: Fordham University Press, 1999); Janet Duitsman Cornelius, *"When I Can Read My Title Clear": Literacy, Slavery, and Religion in the Antebellum South* (Columbia: University of South Carolina Press, 1991); Samuel L. Horst, *Education for Manhood: The Education of Blacks in Virginia during the Civil War* (Lanham, MD: University Press of America, 1987); William A. Link, *A Hard Country and a Lonely Place: Schooling, Society, and Reform in Rural Virginia, 1870–1920* (Chapel Hill: University of North Carolina Press, 1986); Watkins, *White Architects of Black Education;* Williams, *Self-Taught.*

11. William A. Gladstone, *Men of Color* (Gettysburg, PA: Thomas Publications, 1993), 36, 92; Phillips, "Education of Blacks in Tennessee," 98; McPherson, *Negro's Civil War,* 179–92, 315–19.

12. Quarles, *Lincoln and the Negro,* 177.

13. Gladstone, *United States Colored Troops,* 13–16, 39.

14. Breveted a brigadier general during the Civil War, C. F. Adams Jr. was a member of the prominent Adams family (his grandfather was John Quincy Adams). At age twenty-six Adams volunteered for the Union army and was commissioned first lieutenant, Company H, First Massachusetts Volunteer Cavalry. He served with his company as it fought in South Carolina and in the September 1862 Antietam campaign. Promoted to captain and commander of Company H on October 30, 1862, he directed the unit during its participation in the June–July 1863 Gettysburg campaign and fought in the heavy cavalry clash at Aldie, Virginia, on June 17, 1863. Commissioned a lieutenant colonel on July 15, 1864, Adams then received orders to be second in command of the Fifth Massachusetts (Colored) Volunteer Cavalry. The regiment fought in the Siege of Petersburg, Virginia. See Cornish, *Sable Arm,* 132.

15. Cornish, *Sable Arm,* 132.

16. Cornish, *Sable Arm,* 132; see also Glatthaar, *Forged in Battle.*

17. James M. McPherson, *Battle Cry of Freedom: The Civil War Era* (New York: Oxford University Press, 2003), 668; see also Glatthaar, *Forged in Battle,* 123–24.

18. Daniel Ullman quoted in Glatthaar, *Forged in Battle,* 129–30.

19. *New York Times* quotation is in Glatthaar, *Forged in Battle,* 130; see also Duncan, *Where Death and Glory Meet,* 51.

20. Westwood, *Black Troops, White Commanders,* 25–29, 32, 86–89, 92, 103n27. See also US Army War College, *The Use of Negro Manpower in War,* AWC 127–25, October 30, 1925 (Carlisle Barracks, PA: US Army Military History Institute) (hereafter cited as AWC 127–25).

21. James M. McPherson, *For Cause and Comrades: Why Men Fought in the Civil War* (New York: Oxford University Press, 1997), viii–ix, 126–28; Cornish, *Sable Arm,* 105–8, 133; Duncan, *Where Death and Glory Meet,* 51; Gladstone, *United States Colored Troops,* 9. On May 22, 1863, the War Department issued General Order No. 143 establishing the Bureau of Colored Troops. The bureau was responsible for recruiting African American soldiers, commissioning officers to command them, organizing regiments, and maintaining their records. The first regiment of USCT mustered into the federal service in Washington, DC, on June 30, 1863. The last regiment, the 125th, did not muster out of service until December 1867. Fourteen states raised volunteer units under their state designations, which were eventually redesignated as USCT. Almost exclusively, white officers commanded these segregated regiments. Three states raised colored regiments that maintained their state designation: Connecticut, Louisiana, and Massachusetts. By the end of the war, 178,975 enlisted men served in the US Army as members of the USCT. Another 9,695 African American men served in the United States Navy, and by war's end sixteen Black infantrymen had received the Medal of Honor for valor. See Gladstone, *United States Colored Troops,* 11; Smith, *Black Soldiers in Blue,* 8; Gladstone, *Men of Color,* 185.

22. Cornish, *Sable Arm,* 105–8, 133; Duncan, *Where Death and Glory Meet,* 51; Gladstone, *Men of Color,* 185; Gladstone, *United States Colored Troops,* 9–11; Smith, *Black Soldiers in Blue,* 8.

23. "Negro Troops," *Harper's Weekly* 7.338 (June 20, 1863).

24. Glatthaar, *Forged in Battle,* 128; Noah Andre Trudeau, *Like Men of War: Black Troops in the Civil War, 1862–1865* (Boston: Little, Brown, 1998), 26, 27, 34–45, 288, 396, 467.

25. Smith, *Black Soldiers in Blue,* 54; Trudeau, *Like Men of War,* 26, 27, 34–45, 288, 396, 467.

26. Grierson quoted in Smith, *Black Soldiers in Blue,* 55. Grierson was an especially talented Union cavalry commander.

27. "The Siege of Port Hudson: An Official Report from General Banks," *New York Times,* June 10, 1863.

28. "Negro Troops," *Harper's Weekly,* June 20, 1863; see also Duncan, *Where Death and Glory Meet,* 51.

29. Foner, *Reconstruction,* 593; see also Eric Foner, *The Second Founding: How the Civil War and Reconstruction Remade the Constitution* (New York: W. W. Norton, 2019), 195–200.

30. 39th Congress, Session 1, July 28, 1866, https://www.loc.gov/law/help/statutes-at-large /39th-congress.php; see also George Washington Williams, *A History of the Negro Troops in the War of the Rebellion, 1861–1865* (1887; reprint, New York: Fordham University Press, 2012), 139–40; McPherson, *Negro's Civil War,* 179–92, 315–19. For more on the Lost Cause ideology that emerged after the Civil War and a recasting of the struggle to perpetuate slavery as a heroic defense of states' rights, specifically in North Carolina, see Adam H. Domby, *The False Cause: Fraud, Fabrication, and White Supremacy in Confederate Memory* (Charlottesville: University of Virginia Press, 2020).

31. Foner, *Reconstruction,* 231. See also Holly Pinheiro, *The Families' Civil War: Black Soldiers and the Fight for Racial Justice* (Athens: University of Georgia Press, 2022).

32. For more on Port Royal, see Kevin Dougherty, *The Port Royal Experiment: A Case Study in Development* (Starkville: University Press of Mississippi, 2014); Willie Lee Rose, *Rehearsal for Reconstruction: The Port Royal Experiment* (Athens: University of Georgia Press, 1999).

33. See Glatthaar, *Forged in Battle.*

34. Although the 39th Congress authorized six Black regiments, budget cuts led to reductions across the whole army, which whittled the Black regiments down to four (Ninth and Tenth Cavalry, 24th and 25th Infantry). See Robert Wooster, *The United States Army and the Making of America: From Confederation to Empire, 1775–1903* (Lawrence: University Press of Kansas, 2021), 222–40; see also 39th Congress, Session 1, July 28, 1866, https://www.loc.gov/law/help/statutes-at-large/39th-congress.php.

35. Lieut. Henry Ossian Flipper, *The Colored Cadet at West Point: Autobiography of Lieut. Henry Ossian Flipper, U.S.A., First Graduate of Color from the U.S. Military Academy* (New York: Homer Lee & Co., 1878), 178.

36. Tom Carhart, *Barricades: The First African-American West Point Cadets and Their Constant Fight for Survival* (Bloomington, IN: Xlibris US, 2020), 27.

37. Flipper, *Colored Cadet at West Point,* 20–21.

38. Flipper, *Colored Cadet at West Point,* 20–21. These courses focused on the history of the United States, arithmetic, and the various operations in addition, subtraction, multiplication, and division.

39. Flipper, *Colored Cadet at West Point,* 21–22.

40. Kate Masur, "A Rare Phenomenon of Philological Vegetation: The Word 'Contraband' and the Meanings of Emancipation in the United States," *Journal of American History* 93 (March 2007):1050–84; see also Col. Robert Gould Shaw's February 8, 1863, letter to his wife, Annie, in Shaw, *Blue-Eyed Child of Fortune,* 51, 285–86.

41. Charles Sumner Wilson never attended West Point. See Gerald Gill, "Another Light on the Hill: Gerald Hill Introduction" Charles Sumer Wilson Exhibit, Amherst College Archives and Special Collections, https://exhibits.tufts.edu/spotlight /another-light/feature/charles-sumner-wilson.

42. "West Point Revels," *New York Sun,* May 25, 1870; see also Carhart, *Barricades,* 30, 30n7.

43. "West Point Revels," *New York Sun,* May 25, 1870.

44. Carhart, *Barricades,* 30–31, 30n7.

45. Carhart, *Barricades,* 31. Smith was dismissed from USMA in the spring of 1874 following a failure in annual examinations in experimental philosophy.

46. Flipper, *Colored Cadet at West Point,* 29.

47. AWC 127–25.

48. Flipper, *Colored Cadet at West Point,* 321–22.

49. Gladstone, *Men of Color,* 97.

50. Eaton, *Grant, Lincoln, and the Freedmen,* 208.

51. Adjutant General to Titus N. Alexander, February 25, 1898, File 137 0. M.A. 1898, Correspondence Relating to the U.S. Military Academy, 1867–1904, Records of the Adjutant General's Office, 1780's–1917, Record Group 94, National Archives and Records Administration, Washington, DC. For further clarification on this War Department quotation, the West Point archives could shed greater light. See also Carhart, *Barricades,* ix, xii. The document "Statement Showing the Number of Colored Persons Appointed Candidate for Admission to the U.S. Military Academy," October 21, 1886, RG 404, USMA Library Special Collections and Archives proves this statement to be less than truthful.

52. Carhart, *Barricades,* ix–x.

THE FIRST AFRICAN
AMERICAN CADETS
AT WEST POINT

Makonen Campbell and Louisa Koebrich

Although the Civil War resolved questions over Black freedom and military service, battles over what constituted citizenship in American society were just beginning. West Point became a salient battleground on this new front. Between 1870 and the end of Reconstruction in 1877, sixteen Black men were nominated to West Point. They were Charles Sumner Wilson, Henry Alonzo Napier, Michael Howard, James Webster Smith, James Elias Rector, Thomas Van Rensselaer Gibbs, Henry Ossian Flipper, John Washington Williams, William Henry Jarvis Jr, William Henry White, Whitefield McKinlay, William Narcese Werles, Johnson Chestnut Whittaker, Joseph Thomas Dubuclet, John Augustus Simkins, and Charles Augustus Minnie. Of these, eight gained admission, and only one graduated and was commissioned as an officer in the United States Army. This chapter explores their experiences and motivations and also analyzes why they were chosen and how their contributions to integrate the United States Military Academy affected those that followed them.

Collectively, Black cadets' experiences illustrate the ways they tried to navigate and fulfill the promise of opportunity at the academy. These cadets were carefully selected for their aptitude, potential, and preparation. Whether they had begun life enslaved or were only a generation removed from it, as a whole they had already accomplished great feats in education and were well supported by powerful advocates for their success. Their supporters had a strong desire to see them graduate. When each failed or was dismissed, it fulfilled and amplified a warped narrative that these were somehow the "wrong cadets" and that their failures reflected a lack of preparation or fitness rather than any shortcomings elsewhere among the academy, faculty, or corps of cadets. Instead, as one cadet departed, another stepped up, forging a new path for others to follow. They adapted and experimented out of necessity as path after path seemed to lead to dismissal from the academy. They understood their task was unique and that the standard formulae by which white cadets succeeded at West Point did not apply to them. As failure became repetitive, it became almost routine. Always striving to graduate but conscious of the probability of dismissal, Black cadets found ways to leverage their experiences at West Point, however short or long, to their advantage after departing the academy.

These hopeful young men ventured to West Point to change their lives, to celebrate the changes they saw in the nation around them, and to build a new and enduring chapter in the record of Black military service. Instead, the racism that was endemic at all levels of the academy defined strict limitations for those first African American cadets—limitations that in the end proved barriers for nearly all who made the attempt. Although the dismissal of most of the Black cadets who received nominations during the Reconstruction era may have been viewed as the nadir of their lives, many of these cadets went on to have successful careers, and what is more, they and the communities that received them took great pride in their time at West Point regardless of whether they graduated. In this, there is found an element of progress on the national stage, even if the course itself was marred by apparent failures.

Their story is something of a patchwork. There are large gaps in the records and archives, as is the case with so many Black soldiers, especially those who served in the nineteenth century. While each of these men had unique stories and experiences, there are common threads that defined life at West Point for its first Black cadets. Paradoxically, there is a great deal of tragedy to be found in their stories, but there is also great triumph. These men were inspired and inspiring, and through their collective struggle rose a new claim to citizenship and public service—Black officership.

The archival gap is acute. So many of the narratives and experiences of these first Black cadets are lost to the record, and unfortunately, the most accessible stories are those recovered in missteps—namely the investigations, courts-martial, and reports of failures. Those cadets who met with neither exceptional success nor exceptional failure remain largely absent from the record. But the overall arc of their stories is so similar that it is possible to gather a sense of what experiences they might have shared. Taken collectively, these young men entered West Point full of hope and potential. They had been selected because they seemed well positioned to succeed.

Most struggled academically, but even those who performed well academically tended to struggle in their military duties, particularly in their first six

months. Any cadet who accumulated ninety-nine or more demerits in a semester would be discharged, and although the first Black cadets tended to ac-
cumulate demerits rapidly, none were dismissed for reaching or exceeding that
threshold number. Still, some white cadets were dismissed for reaching that
threshold during the same periods, which indicates that while Black cadets
may have received more demerits than most white counterparts, they were not
targeted this way to the point of dismissal.[1]

A deeper inspection also reveals just how isolated these cadets were. The
fact that they were not dismissed while some white cadets were suggests their
demerits resulted not just from a lack of knowledge of drill and military cus-
toms but also from a lack of shared or communal learning—that system of informal mutual support that marked the cadet lives of their white counterparts.
Very few cadets arrived at West Point knowing how to drill or pass inspections.
They learned through the class system, from the mentorship of senior cadets,
and, perhaps most importantly, from the daily trials and tribulations of one another. While not every white cadet unfairly targeted their Black counterparts,
none would break ranks to go so far as to help them outside of the official line
of duty. After the first semesters, those Black cadets who remained enrolled
saw marked drops in their numbers of demerits, demonstrating that they could
and did adapt, albeit at a slower rate than the average white cadet. That the
Black cadets took longer to adapt illustrates how central fellow cadets were to
any single cadet's learning process—a point that reveals how isolation posed a
real threat to cadets' ability to remain at West Point.

Black cadets' isolation and slower pace of adjustment were products of the
practice of silencing, sometimes referred to as "cutting." Silencing was the practice of ceasing any recognition of a cadet whom the broader corps of cadets
perceived to be guilty of an unacceptable transgression. Although officers and
professors appeared professional in their day-to-day interactions with African
American cadets, as a group they implicitly approved of the silencing of Black
cadets.[2] For the silenced—including every Black cadet—social interactions were
permitted only during the conduct of official duties, as in classroom instruction
and drills. Beyond official duties, Black cadets were isolated, threatened, and
denigrated by their white counterparts. This had a direct effect on their ability
to learn and thrive and would take a heavy toll on them mentally, requiring exceptional resilience to endure and succeed where their white counterparts could
share burdens and lessons learned with friends and classmates in order to survive.

When Radical Republicans such as Representatives Benjamin Butler (Massachusetts), Legrand W. Perce (Mississippi), Solomon L. Hoge (South Carolina), and Robert Smalls (South Carolina) appointed young Black men to

West Point, they understood that they were attempting to reshape the social fabric of the academy—opening a new chapter in democratizing the institution. The academy had long been charged with elitism when it was supposed to represent the entirety of the nation. These new Black candidates would encompass every walk of life, and through military service prove their mettle while serving as representatives for an entire race. Their efforts were part of a larger agenda to reshape public ideas about the meaning of Black citizenship in the United States. According to one historian, "Benjamin Butler [sought] a young African American man to accept an appointment" knowing that the nominee would have to face "insults, taunts, and social exclusion."[3] Republicans searched for exceptional African American candidates for five years before finally settling on Charles Sumner Wilson for the initial appointment.

In the spring of 1870, a young doe-eyed Charles Sumner Wilson stood in the home of his mother, Rebecca Wilson, and read the nomination letter that was sent to the secretary of war. It was a standard army form that acknowledged Wilson's age and place of residence and was signed by Representative Benjamin Butler of Massachusetts. What was unsaid in the letter was that Butler believed that Wilson was fully qualified to attend the academy despite his age (he was not yet seventeen) and that he intended to push this nomination through the confirmation process. His nomination would mark the first appointment of an African American to the academy by a sitting congressman.[4] It also marked the beginning of a struggle to integrate African Americans into West Point.

Butler had been looking for a fitting African American candidate from Massachusetts to open the door for opportunities for African Americans at the academy. He was aware that Rebecca Eldridge Wilson, a prominent figure within the African American community, had a son. Butler may have been familiar with the family, and Wilson's mother specifically, through her work as a member of the Salem Female Anti-Slavery Society before the Civil War or as a widow of the war afterward.[5] Admittedly, there is little else known about her son Charles or why he was selected to be the first African American nominee to West Point. The evidence indicates that Butler viewed Wilson's pedigree as fitting enough to measure up the other candidates entering West Point that summer. His father served and died, fighting with the all-Black 55th Massachusetts Infantry Regiment in the Civil War. Wilson also graduated at the top of his class in Salem.[6] Many of the white candidates nominated were sons of current or former army officers as well. Wilson was well qualified, and if he had not been so young—he had not yet reached the required age of seventeen when he was nominated for his appointment—he might very well have passed the entrance examination and become the first African American to attend the academy. Denied entry due to his age, Wilson went on to attend Tufts University and Amherst College before becoming the first African American lawyer in Essex County, Massachusetts.[7]

The African American cadets who arrived at the academy had to contend with assertions of inferiority, suggestions of preferential treatment, and racist backlash to their arrival. In June 1870 Michael Howard and James Webster Smith arrived at West Point with the weight of the world on their shoulders. Neither Howard nor Smith had an agenda when he arrived, and neither espoused anything that would be considered radical ideas about social equality. However, their arrival did not sit well with the faculty, cadets, and prospective cadets alike. The African American cadet candidates who arrived were viewed as an aberration. Observing from outside the academy, Nelson A. Miles—a noted Civil War commander and future commanding general of the US Army—spoke for many others in the officer corps when he argued that the "ages of slavery had reduced [African Americans] to the lowest ebb of manhood" and thus left them ill-suited for the mental rigors of life as a cadet.[8] The consensus of the corps was that white cadets could be forced to drill with Black cadets, "but they cannot be made to associate with [them]."[9]

On May 25, 1870, Michael Howard was the first African American to arrive at West Point as a cadet candidate. Born in Fayette, Mississippi, on a cotton plantation where he spent a portion of his early life enslaved, he was also enrolled in a freedman's school for the three years preceding his nomination to West Point. While it is impossible to assess his academic prowess prior to the entrance exams at West Point, he was certainly a compelling candidate to Representative Legrand Perce, for Howard was literate at a time when most of the Black population in Mississippi was not. Upon reporting to the commandant's office, he proudly presented his papers but was turned away because he had arrived well in advance of the examination period. After being turned away, Howard sought to secure lodging until the examination period began. He quickly discovered that there were no lodgings available on post, nor in Highland Falls, the town immediately outside West Point. The lack of accommodations may have been more the result of bad timing rather than a racist response to his arrival at West Point. Those who had arrived to observe the annual examinations likely secured all guest lodging on the post. He finally found relief with a local African American family, who entertained him and provided lodging until the examination period began.[10] Howard's introduction to West Point was representative of so many of the routine moments these first Black cadets would experience. Neither overtly racist nor implausible, these vignettes nevertheless combined to create a tableau in which these Black cadets were made to feel unwelcome—at least in white society—but more likely made to feel disparaged, disrespected, and disheartened.

James Webster Smith arrived shortly after Howard. Smith, who would become the first Black cadet admitted to the academy, hailed from Columbia, South Carolina, He was the son of a formerly enslaved African American family. His father served in the army during the Civil War and fought under Sherman's command. After the war Smith's father continued to work as a carpenter and earned a decent living that allowed him to provide for his family and send his son to a freedmen's school in Columbia. Smith's mother, with her fair complexion, was considered an "octoroon" (or one-eighth Black) and had "received an education from a private school through her white father's influence." Understandably, then, Smith received the best possible education, especially compared to other children of formerly enslaved couples. Smith was an excellent student, and his academic performance caught the attention of David Clark, a philanthropist and Civil War veteran himself. Clark would serve as Smith's benefactor, taking Smith on as his ward and escorting him to Hartford, Connecticut, to further his education.[11]

By the time of his appointment to West Point, Smith had graduated at the top of his class from Hartford High School and was enrolled at Howard University. General O. O. Howard reassured Clark that Smith would receive a fair chance at the academy. Clark later recalled, "I was averse to the idea of his going because I thought I could foresee the difficulties he would encounter at West Point on account of his color, but I left the whole matter with Gen. Howard, and he favored the proposition, Smith accepted, and went to West Point."[12] General Howard and all of the Republicans who nominated African American cadets to attend West Point anticipated that those cadets would face hardship that would require them "to possess extraordinary qualities."[13] Smith was prepared for the academic rigors of attending the academy but would soon be tested socially as he navigated an environment that was wholly unprepared for African American cadets.

Smith's story is remarkable for so many reasons. Though there are limited sources on his time at West Point, it is clear that he had a vibrant personality. He approached West Point in a self-assured manner, confident of his place. He had, after all, been assured by his father—a veteran under General Sherman's command—that he would be treated fairly as a cadet. Smith's introduction to how he would be received at the academy occurred when he entered the Rose Hotel and asked to be seated and fed by the clerk. The clerk responded, "A meal of victuals for a nigger? Well, you'll have to be hungry for a good while if you wait to get something to eat here."[14] The clerk's attitude was indicative of the general mood at the academy and did not bode well for the notion that Smith would be treated fairly, as his father had promised. Smith noted that he "had not been there an hour before [he] was reminded by several thoughtful cadets that [he] was nothing but a damned nigger."[15] Even so, Smith's spirit and sense of self-worth were not so easily broken. His first year was riddled with overt acts of hostility, and he stood up for himself not just out of pride but with a sincere expectation that the academy—and the nation—would do better.

Before the entrance examinations, Smith and Howard roomed together and were able to study together when not at drill. The outlook on Howard's chances of passing the exam was low. He had only recently begun to receive a formal education. Knowledge of Howard's dearth of formal education coupled with the negative perceptions of Howard's appearance (and an accent that observers thought "smacked of the plantation") shaped the opinions of the "aristocratic professors and jaunty cadets" eager to see him fail his entrance examination.[16] Both his father and Representative Perce wrote letters to the Academic Board requesting that Howard's entrance examinations be postponed until he could be properly prepared both financially and academically for the rigors of life at the academy, but to no avail.[17]

During their brief shared experience of West Point, several grievous incidents tested Howard's and Smith's resolve, made it difficult for them to remain focused on passing the entrance examinations, and foreshadowed life after admission. While sleeping in the middle of the night, Howard and Smith were doused with waste and excrement from a slop pail. Smith's and Howard's report of the incident generated a half-hearted investigation that identified no culprits, and nothing further came of the event.[18] Most of our knowledge of these types of events comes from investigation reports, which necessarily means that any incident not reported or investigated is absent from the record. It is impossible to say what other indignities these men might have suffered, though in Smith's first year, there is every indication that he was a strong self-advocate and likely reported the most egregious transgressions. Even so, this incident set a clear precedent: there was a great deal of room for abuse under the guise of anonymity. A second incident occurred at the bootblack's shop. Cadets were lined up to have their shoes shined when Robert McChord—a prospective cadet from Kentucky—shoved Michael Howard, telling him to move out of the doorway. McChord later slapped Howard for still not moving out of his way and then threatened to "cut him open."[19] Though Howard was the primary target of the abuse and was certainly upset, Smith was the one who pushed an investigation, going so far as to write Howard's statement at Howard's request. As there were slight but insignificant discrepancies between the written account and the later questioning, the investigators determined this to be an exaggerated account. McChord was admonished, but nothing more came of the incident.[20]

Smith's troubles compounded when the academic board dismissed nearly 53 percent of those appointed to enter West Point that summer, including Michael Howard.[21] As West Point's first and only admitted Black cadet, Smith would face a tumultuous year isolated at the academy, during which he would suffer incredible and outrageous indignities. Over the course of Smith's first year, he would be subject to two courts-martial. The first revolved around two back-to-back incidents. As a new cadet, Smith was under considerable stress and unsure of rules. His difficulties as a plebe that summer were compounded by his isolation and the sense of mutual support denied to him because of his color. In his duress, he and another cadet came to blows, though it is unclear who actually initiated the fight. Shortly thereafter, Smith contested a demerit report over the incident, which two white cadets insisted had taken place. In the course of the court-martial, the fight was the lesser of two evils compared to charges of lying. The court found that the event in question did take place, though not on the day the report indicated. As a result, Smith was acquitted, but his honor was impugned as everyone from cadets to faculty believed that Smith had lied in essence if not in fact.[22] It is entirely possible that Smith did take advantage of the reporting error to avoid one demerit, but it is equally plausible that with all of the stressors and daily harassment, he had genuinely forgotten the demerit report.

Smith found himself subject to a second court-martial for a substantially similar charge. In the second case, he received a demerit for inattention in the ranks. Smith offered the explanation that the cadet next to him had stepped on his foot. When the cadet denied it, Smith faced new charges for lying. This time, he was found guilty, and while the court decided to dismiss Smith, the secretary of war reduced the punishment to having Smith repeat his first year.[23]

The collective impact of Smith's courts-martial reframed the environment for Black cadets at West Point during Reconstruction. The corps of cadets learned that the Grant administration and its congressional allies were determined to sustain Black cadets at West Point, and the outright acts of aggression would be futile in having him dismissed politically. Instead, they would have to isolate him and hope he failed, which became increasingly likely when he did not have access to the same resources and support the corps found in each other.[24] For Smith, the lesson he sadly internalized was that after a year of standing up for himself against injustice, it would always come down to his word against that of several white cadets. And with no one to stand on his side, he would always be considered the offending party.[25] Smith concluded that his survival at West Point demanded quiet forbearance rather than challenging the injustices he experienced. Surviving records after his crowded first year are tellingly scant. He was the subject of no more major investigations and few news reports until after he was dismissed from the academy.

As Smith repeated his first year in 1871, Henry Alonzo Napier of Tennessee reported. Napier passed his entrance exams and spent an entire year with Smith before failing exams and being dismissed from the academy. Napier was nominated by W. F. Prosser, a US Army veteran of the Civil War and a Republican elected to represent Tennessee in Congress in 1869. Napier and Smith both attended Howard University for a time and were contemporaries while there. Smith had actually hoped to "delay his admission to West Point until both he and Napier could enter and graduate," indicating an understanding that attending the academy would be an endeavor best experienced as a team, even if only of two.[26] Faculty commented on Napier's deportment, stating that he "is dignified, complaiant [*sic*], ready spoken, and quite charming as a conversationalist. He is said to be very patient and even-tempered, but not at all likely to suffer being trampled upon unjustly." By all accounts, Napier was "of a thorough education, manly qualities, and devoted duty," characteristics that suggest he was constitutionally well suited for the US Military Academy.[27] Over the course of their year together, Smith undoubtedly coached Napier in how to succeed at West Point. While they likely did not suffer the kind of outrageous incidents such as those Smith first encountered with a slop pail in the night, they continued to experience silencing and, most likely, hostile acts short of those requiring investigation. For their parts, the fact that no investigations were initiated likely means that Napier, under Smith's mentorship, may not have reported harassment or tended to avoid the worst of its intensity. Smith warned Henry O. Flipper to avoid controversy, and there is no reason to expect he did not communicate a similar warning to Napier.[28] Over the course of that year Smith performed well in classes he repeated from his first year, and Napier followed a pattern most of these first Black cadets experienced. He had a high number of demerits over that first year, but, unlike Smith, Napier's demerits never improved, which suggests that Napier had a harder time adapting to the discipline of West Point, even with Smith there as a guide.[29]

Napier ultimately failed his examinations in June 1872 and was dismissed for academic deficiency. The record notes that faculty members reported that Napier was temperamentally unsuited for the academy, having been ill-disciplined, inattentive to regulations, and frivolous. If a cadet failed an exam, they were then evaluated by faculty to determine if there was a reason to give them a chance to repeat the year. These reports categorized each cadet with consistent descriptors ranging from "very inattentive" to "very attentive" to regulations, discipline, studiousness, and other qualities. While the consistent language and categories of evaluation gave a veneer of objectivity, it is clear where there was room for bias to enter the assessment. Nine cadets failed exams that year, including Napier, and three were afforded the opportunity to remain. The record shows no sign of outright racism because only those cadets who received the best reports in these pseudo-scientific evaluations were given the chance to remain, and that is consistent across the years with white cadets as well. Thus, anytime a superintendent investigated the reports for irregularities, he could conclude that nothing untoward had happened and defer to the judgment of the faculty. Napier was the only cadet who was identified as having "very little aptitude" and being "very inattentive to regulations," and he was dismissed from the academy.[30] While there were very clear flaws in the institutional approach, it is equally undeniable that Napier struggled academically and did not show the same improvement in discipline that Smith did. His dismissal also reinforced the realization that academic failure was the surest way to see Black cadets out of the academy after Smith had endured and survived a year's worth of courts-martial.

As Smith began his third year, James Elias Rector and Thomas Van Rensselaer Gibbs reported to West Point. Rector failed the entrance examination and did not gain admission. But one month after Napier departed, Gibbs gained admission and joined Smith at West Point. As with so many of these first Black cadets, Gibbs had excellent credentials. His father, Jonathan Gibbs, served as Florida's secretary of state. Continuing the pattern established by Smith and Napier, Gibbs attended Howard University prior to entering West Point and was seemingly well prepared for the academy.

Nevertheless, Gibbs survived for only six months before midyear examinations in January 1873. A family friend wrote to the superintendent requesting an investigation because young Cadet Gibbs had reported that injustice had been done—most likely in reference to his recent exam failures. Superintendent Ruger responded that an investigation revealed no irregularities in the examinations. He further explained that while Gibbs could be renominated, the Academic Board advised against it because its members believed Gibbs lacked the aptitude to succeed. Otherwise, Ruger said that Gibbs had been well liked, which was an interesting observation that indirectly contrasted Gibbs with Smith, who was far from well liked yet managed to continue to scrape by in his exams. The fact that the branded liar continued while the more likable Black cadet was dismissed suggests the exams themselves were at least administered impartially. But perhaps more significant is the extent to which Gibbs, Gibbs's father, and their acquaintances resisted the dismissal. Whatever the experiences at West Point, Gibbs wanted to be there, and it was important to more people than just him.

In 1873 two new Black cadets reported and passed the entrance examinations to gain admission—and one would become West Point's first Black graduate. Henry Ossian Flipper of Georgia and John Washington Williams of Virginia joined Smith at West Point. Perhaps by this time, Smith had begun to feel that this was his journey to take alone, as no one besides himself had lasted more than a year. Upon Flipper's arrival, Smith had sent him a letter advising him on what to expect and suggesting how he might best get along at the academy, the implication being that Flipper should not seek justice as strongly as Smith had but rather go with the grain. Smith said that Flipper need not be afraid of "blows or insults" and advised him to "avoid any forward conduct" if he wished to "avoid certain consequences." Flipper said of the letter that nothing had "so affected me or influenced my conduct at West Point as its melancholy tone."[31] It is possible that Smith offered similar guidance to other African American cadets as they arrived at West Point. This suggests the extent to which each of these cadets relied on those who came before to gauge the best possible paths to success, or at least to gain general ideas on how to avoid failure. Smith's advice seems to have helped Williams in particular, who had fewer demerits in the fall of 1873 than Flipper, indicating that he adapted better and more rapidly to the discipline and rigor of military drill. In January 1874, however, Williams failed his French exam and was dismissed. His representative asked for him to be reappointed as he had never studied French before. As there were no irregularities in the exams, the board saw no reason to give Williams another chance. Undoubtedly, there were white cadets who had not had the chance to study French before, and thus to the board this would not have seemed like a sufficient reason to make an exception. Of the ten cadets dismissed that semester, none were offered a second chance.[32]

After four years at West Point, James Webster Smith was found deficient in his natural and experimental philosophy examination and was dismissed from the academy. He once more appeared in the public discourse as he wrote to several national papers, including Frederick Douglass's own *New National Era,* to share how that final examination had been administered unfairly by Peter S. Michie.[33] What is striking about his loud return to public discourse is that he clearly had not been cowed by his experiences at West Point but merely had been adapting. Even more so than Gibbs, Smith fought his dismissal through the court of public opinion, but to no avail. Despite his ostracizing from the corps and lack of consistent companions to share the load, Smith fought with all he had to try to remain and see graduation. He was a truly resilient character to have endured so long and to have shepherded so many into the academy.

The year that Smith was dismissed from West Point, four Black cadets were nominated. Unfortunately, none of them passed examinations. Representative

D. W. Gooch of Massachusetts, who had appointed one of the cadets in 1874 (young William Henry Jarvis Jr.), initiated an inquiry, claiming that Jarvis had been mistreated "by the entire Academy" due to his race, claiming that if Jarvis failed, it would be "because his courage is broken by the treatment." The superintendent investigated, and the results were quite telling. The superintendent asked Jarvis directly whether anyone had insulted or injured him, which put Jarvis in a tough position. As other Black cadets had learned through hard experience, answering affirmatively would undoubtedly lead to an investigation that no white cadet would corroborate, thus putting Jarvis in the position of being branded a liar—similar to James W. Smith and completely incompatible with the values of West Point. Instead, Jarvis answered as he had to: that he had heard comments, other cadets frequently referring to him as "the negro or nigger or that is the darkey [*sic*]," but these comments did not impact his examination. The superintendent was satisfied with this investigation because he considered Jarvis to be "so nearly white that many would not observe him the cause probably of some remarks not designed for his hearing."[34] In other words, because he could pass for white, Jarvis couldn't possibly be upset; therefore, no one could think the comments were directed at him specifically. According to that logic, Jarvis had no cause to be offended directly or suffer second-hand shame. Such responses offer a small window into the pervasive abuse that Black cadets lived with and were expected to adapt to.

It would be two years before another Black cadet gained admission. William Henry White, Whitefield McKinlay, and William Narcese Werles were all nominated prior to Johnson C. Whittaker's arrival, but they were denied entry for various reasons, making Flipper and Whittaker the sole representatives of their race in the corps of cadets in 1876–77. On June 13, the evening before the Class of 1877 would graduate, the revelry of the corps of cadets included a minstrel show.[35] The show possibly included a rendition of a song written by a white cadet that was frequently sung during Flipper's time at West Point, titled "Nigger Jim" (typically performed in blackface). According to one member who remembered the doggerel clearly decades later, "It was aimed at Jim Smith, the colored cadet, a repulsive looking, freckle faced negro."[36] Nevertheless, on June 14, 1877—the army's 102nd birthday, Henry O. Flipper would obtain the ever elusive "sheepskin" that was coveted by the other African American cadets who had come before him. Of his performance during final exams, Flipper said, "My hopes were never higher; I knew I would graduate. I felt it, and I made one last effort for rank." Such simple but powerful words can barely touch the depth of feeling Flipper experienced as he approached a certain graduation. When Flipper received his commission there were cheers from the stands, led by none other than Gen. William T. Sherman.

Although he did not seek to change the social atmosphere of the time, his graduation and commissioning into the US Army were viewed as a "victory not only for himself but for the race."[37]

One year later, Flipper published his memoirs, which serve as one of the greatest sources on what contemporary life at West Point was like. His memoirs frequently downplay the level of abuse and mistreatment that were constant companions during his time there, which only helped sustain the veneer that the institution had treated the cadets fairly. Even in his attempts to downplay the harshness of his experiences, the reality shines through: "Why I was the happiest man in the institution, except when I'd get brooding over my loneliness."[38] The extent to which he was silenced pervaded and impacted the entirety of his experiences there. It is worth remembering, though, that Flipper, one year after graduation, was still living high on the joy of having succeeded where so many had failed. Moreover, he was quite hopeful for and possibly protective of his roommate remaining at the academy and any who might follow him, which most certainly flavored his account. Johnson C. Whittaker was now alone and isolated and would remain so until the arrival of Charles Augustus Minnie in 1877, the last of the Black cadets nominated to the academy during Reconstruction.

There was an opinion that the academy "was not responsible for the paucity of Black cadets" and that the lack of representation was due more to a dearth of fully qualified candidates.[39] This, to some, would explain why so many African American cadets were being found academically deficient and dismissed, although many of the cadets who arrived at the academy had been well prepared through formal education. Gen. John Schofield, in an 1880 report to the secretary of the army, made it clear what his thoughts were on the nomination of African American cadets to the academy. He placed the blame for the lack of proper and fair treatment at the academy squarely at Black feet, stating that "young men so recently emerged from a state of slavery" could not be expected to compete with "those who have inherited the strength in the many generations of freedom enjoyed by their ancestors." He believed that Congress was retarding the education of its future military leaders by "forcing cadets into official positions for which they have not become duly qualified."[40] While Schofield did not necessarily represent the views of everyone at West Point, it is easy to imagine the range of reactions Black cadets faced from outright hostility to quiet antipathy, all underscored by an expectation of eventual failure. Those reactions are explored further in Rory McGovern's contribution to this volume—the important point here is that reactions such as Schofield's shaped Black cadets' experiences in profound ways.

Charles Augustus Minnie of New York arrived at the academy in 1877, providing Whittaker a brief respite from being the only African American at the academy. Minnie's nomination was a result of a competition held in New York's fifth congressional district, represented by Nicholas Muller. Minnie graduated at the top of his secondary class and was reported to be "a wide awake and sharp boy, always attentive to his business, [who] will no doubt make a good army officer in time." Minnie's initial reception may have been affected by his complexion, of which a "slight dusky hue" may have been the only thing that "that would suggest that he was not of the white race."[41] Whether it was because he was not from the Deep South or because he could pass for white, he was surprised by the hostility he encountered at West Point. When interviewed by reporters interested in his experience, he readily provided insight into his treatment, stating that it was "sickening" and that had he been used to the hardships and ill-treatment that his southern counterparts had been accustomed to, he may have been able to endure.[42] The daily pain and indignation that he suffered became unbearable after only six months at the academy. He was found deficient in mathematics and dismissed. Minnie professed to have been perfectly able to pass the math examination but explained that he no longer had the motivation to remain at the academy under those conditions. Whether in reality he could have passed the exam or not, the treatment clearly weighed heavily on him.

This left Whittaker again as the sole African American cadet at the academy until he, too, was dismissed. For nearly four years Whittaker endured, though his academic performance was slowly slipping. Shortly before year-end exams in 1880, Whittaker was found lying on the floor of his room, bound, bleeding, and unconscious. Schofield began an inquiry that came to the conclusion that he and a number of other officers had jumped to on the day of the assault—that Whittaker had masterminded his own attack in an effort to avoid taking an examination he anticipated failing. While the investigation was ongoing and Schofield deliberated the next steps, Whittaker took his year-end examinations and failed natural and experimental philosophy—the same class and examination that caused James W. Smith's dismissal in 1874. Before any action could be taken, the inquiry was suspended and a court-martial began—a process that would take over two years to complete.[43]

While the court-martial ultimately found Whittaker guilty, the conviction was overturned. In the end, Whittaker was still dismissed from the academy because he had failed the June exam two years prior. He was not offered an opportunity to retake it or start the year over. Given Smith's experiences in courts-martial and the hard lessons he shared about justice for Black cadets, it is difficult to imagine that Whittaker would have concocted such a scheme. There had been only one successful cadet, and Whittaker had observed first-hand that Flipper succeeded by staying out of scrutiny.[44] Whittaker departed

West Point bitter about his treatment but resolved to the fact that "I have an education which none can take away from me."[45]

There is a persisting historical narrative that all of the Black cadets and cadet candidates during Reconstruction were ill-prepared, socially, academically, and morally. The reality is that these first Black cadets were carefully selected as political projects whose nominating representatives desired to see them succeed. These young men had as much preparation and potential as could be found in the Reconstruction era. Furthermore, they had the will and motivation to succeed. As anyone who has taught can attest, will and resilience account for so much more of a person's likelihood to succeed than any natural talent. This realization in turn speaks to just how adverse the conditions at West Point were for these first Black cadets. Their stories demonstrate the exceptional resilience required of them and show that such resilience would still not be enough to guarantee success at the academy. Their experiences recast ideas of success and failure. The ultimate marker of success—reaching graduation—was no guarantee of a successful army career, but neither did failing out of the academy mean social failure. Despite leaving short of graduation, they would continue to be known in their communities as West Point men.

While James Webster Smith may not have left West Point with a commission, his time at the academy rendered him an expert in the eyes of his community, and he became the commandant of cadets at South Carolina Agricultural and Mechanical Institute.[46] Johnson Whittaker studied for and passed the bar and practiced law for many years in South Carolina, no doubt motivated by his own travesty of justice. Following that he taught at all-Black schools in Oklahoma and South Carolina, passing on his expertise not only to his students but also to his sons, who both served in World War I. He was known throughout the Black community of Sumter, South Carolina, as "the famous ex-West Point cadet," who was a strict disciplinarian and was quite well respected.[47] Regardless of the circumstances under which these first Black cadets departed, their communities recognized their time there as bestowing them with a level of achievement deserving recognition.

What, then, can be made of the experiences of these first Black cadets during Reconstruction? By all accounts, the environment that the cadets entered was hostile, making it difficult for them to succeed at the academy. Professor Michie maintained that "the isolation had nothing to do with hatred" and pointed to the investigation of Whittaker's assault as proof that the academy treated its African American cadets fairly. He also stubbornly argued that "in no case has there ever been the slightest indignity ever offered to" African American cadets.[48]

In the decade immediately following Michie's comment (1890s), another eleven Black men would be nominated and two would graduate before the doors seemed to firmly close for Black men for another forty-seven years. In that regard, the hard-fought lessons could not be passed on to future generations. But it is equally undeniable that these successes inspired those that followed. Their influence is evidenced by statements such as the one made by John Hanks Alexander (the next Black graduate after Flipper), who remarked that he was inspired to attend West Point because of the "other colored boys who have distinguished themselves there."[49] He did not single out Flipper, then the lone graduate, but instead was inspired by the shared struggles of each Black cadet. In subsequent historical studies, it became only natural to focus on only those that reached graduation, but the unglamorous yet heroic reality is that these collective experiences—including both success and failure—forged realities at West Point and sounded the clarion call for more to follow.

Notes

1. Consolidated Weekly Class Reports and Conduct Rolls of US Military Academy Cadets 1867–94, National Archives Record Group 94, United States Military Academy Library Special Collections and Archives (hereafter cited as USMA), M1002, Roll 2.

2. John F. Marszalek, *Assault at West Point: The Court-Martial of Johnson C. Whittaker* (New York: Scribner's, 1972), 19. For more on the faculty's response to Black cadets, see Rory McGovern's "You Need Not Think You Are on an Equality with Your Classmates," in this volume.

3. Theodore J. Crackel, *West Point: A Bicentennial History* (Lawrence: University Press of Kansas, 2002), 145.

4. Cat Rosch, "Charles Sumner Wilson," "Another Light on the Hill: Black Students at Tufts," https://exhibits.tufts.edu/spotlight/another-light/feature/charles-sumner-wilson.

5. Rosch, "Charles Sumner Wilson."

6. Katherine Whittemore, "The Black Men of Amherst Left Out: Untold Histories of Black Alumni," https://www.amherst.edu/amherst-story/magazine/issues/2021-spring/bicentennial/the-men-black-men-of-amherst-left-out.

7. Whittemore, "The Black Men of Amherst Left Out."

8. Nelson A. Miles quoted in Theophilus G. Steward, *The Colored Regulars in the United States Army 1904* (New York: Arno Press, 1969), 5–6.

9. "West Point: The Opening Seasons, Military and Otherwise," *New York Herald*, May 31, 1870.

10. "West Point: The Opening Seasons, Military and Otherwise," *New York Herald*, May 31, 1870.

11. For more on Smith, see Rory McGovern, Makonen Campbell, and Louisa Koebrich, "'I Hope to Have Justice Done Me or I Can't Get Along Here': James Webster Smith and West Point," *Journal of Military History* 87.4 (October 2023): 964–1003.

12. David A. Clark to Sayles J. Bowen, July 23, 1872, reprinted in "Grant vs. Smith," *New York Tribune,* July 31, 1872. O. O. Howard most likely prompted Hoge to nominate Smith—see McGovern, Campbell, and Koebrich, "I Hope to Have Justice Done Me," 972. For more details on Hoge and Smith's nomination—and the nominations of Black cadets more generally—refer to Adam Domby's essay in this volume, "A Nursery of Treason Remade?"

13. Marszalek, *Assault at West Point,* 18.

14. Henry O. Flipper, *The Colored Cadet at West Point: Autobiography of Lieut. Henry Ossian Flipper, U.S.A., First Graduate of Color from the U.S. Military Academy* (New York: Homer Lee & Co., 1878), 312–13.

15. Flipper, *Colored Cadet at West Point,* 312–13.

16. "The West Point Revels: The Colored Cadet Making Bow to Colleges of the Regular Army—The Consternation of the Caucasian Snobs—A Council of Was—A Plot to Trip Him in His Examination," *New York Sun,* May 25, 1870.

17. Michael Howard Sr. to General Abraham, May 20, 1870; L. Perce, Renomination Letter of Michael Howard, April 13, 1871; and Extract from the Proceedings of the Academic Board, April 26 1871; all in Selected Documents Relating to Blacks Nominated for Appointment to the United States Military Academy during the Nineteenth Century, 1870–1887 (hereafter cited as M1002), USMA.

18. Testimony of James W. Smith, Untitled Transcript of Court of Inquiry addressed to Thomas M. Vincent, AAG, p. 11, James W. Smith Files, M1002, USMA.

19. See "Report of Difficulty between New Cadets Michael Howard and Rob't C. McChord, U.S.M.A.," June 17, 1870, Michael Howard Files, M1002, USMA. The report includes several attached files. Quotations are from "Statement [of Michael Howard], June 6, 1870," and A. Clarke, "Statements of Cadets," both of which are appended to the report. See also Michael Howard to Adelbert Ames, June 9, 1870, Michael Howard Files, M1002, USMA.

20. Testimony of Henry M. Black, Untitled Transcript of Court of Inquiry addressed to Thomas M. Vincent, AAG, p. 44, James W. Smith Files, M1002, USMA.

21. McGovern, Campbell, and Koebrich, "I Hope to Have Justice Done Me," 981–82.

22. Testimony of James W. Smith, Untitled Transcript of Court of Inquiry addressed to Thomas M. Vincent, AAG, James W. Smith Files, M1002, USMA; Unsigned Memorandum from Secretary of War, William Belknap, Undated, James W. Smith Files, M1002, USMA.

23. General Court-Martial Orders No. 6, June 18, 1871, James W. Smith Files, M1002, USMA.

24. McGovern, Campbell, and Koebrich, "I Hope to Have Justice Done Me," 1001–3.

25. See, for example, a letter Smith wrote early during his time at West Point, in which he realized that he could not find white cadets willing to speak on his behalf. James W. Smith to "Friend" [David A. Clark], June 29, 1870, included as an exhibit appended to Untitled Transcript of Court of Inquiry addressed to Thomas M. Vincent, AAG, James W. Smith Files, M1002, USMA.

26. See Pearl, Untitled Newspaper Article, May 25, 1872, Henry Alonzo Napier Files, M1002, USMA. The article places Napier and Smith at Howard University at the same time. Smith endeavored to delay his entry into the academy until both he and Napier could attend.

27. Peter S. Michie, "Caste at West Point," *North American Review* 130.283 (June 1880): 604.

28. Flipper, *Colored Cadet at West Point*, 37.

29. While there are potential alternative explanations for Napier's high number of demerits, the most likely explanation is that he did not adapt well. If there was a concerted and effective effort to target Black cadets through demerits, Smith would have certainly been caught in such a scheme. Cadets reviled Smith for seemingly getting away with being a known liar, the highest offense a cadet could face. Therefore, it seems that Smith managed to adapt in ways that Napier did not. Consolidated Weekly Class Reports and Conduct Rolls of US Military Academy Cadets 1867–94, Henry A. Napier Files, Roll 2, M1002, USMA.

30. Consolidated Weekly Class Reports and Conduct Rolls of US Military Academy Cadets 1867–94, Henry A. Napier Files, Roll 2, M1002, USMA.

31. Flipper, *Colored Cadet at West Point*, 37.

32. Academic Board January 1874 Exam Results, John Washington Williams Files, M1002, USMA.

33. The records at West Point offer no evidence that could corroborate Smith's version of events, which is not to say that his version is not credible but that it cannot be proven in archival records. Many of Smith's letters to various editors are reprinted in full in Flipper, *Colored Cadet at West Point*, 289–309.

34. Telegram from Superintendent Ruger to Secretary of War Belknap, May 1874, William Henry Jarvis Jr. Files, M1002, USMA.

35. Tom Carhart, *Barricades: The First African-American West Point Cadets and Their Constant Fight for Survival* (Bloomington, IN: Xlibris US, 2020), 60.

36. Unpublished Memoir, Eben Swift Files, USMA. There is no evidence that the song was sung while Smith was a cadet there, though it is possible. Swift spoke of singing the song frequently and was a cadet from the years 1872–76.

37. Flipper, *Colored Cadet at West Point*, 239, 253.

38. Flipper, *Colored Cadet at West Point*, 249.

39. Marszalek, *Assault at West Point*, 19.

40. US War Department, *Annual Report of the Secretary of War*, vol. 1 (Washington, DC: Government Printing Office, 1880), 229.

41. Lansingburgh Historical Society, "Charles A. Minnie (b. abt 1848) one of the first African-Americans at West Point," untitled extract from *Saratoga Sentinel,* September 6, 1877, https://lansingburghhistoricalsocietyarchives.org/charles-a-minnie-b-abt-1858-one-of-the-first-african-americans-at-west-point/.

42. "Colored Cadet Minnie: His Return to This City—What He Says about the Daily Life of a Colored Cadet at West Point—His Intention of Joining the City College and Studying Law," *New York Times,* January 26, 1878.

43. Carhart, *Barricades,* 68–69.

44. Flipper has unfairly been described as having not resisted his treatment, which he even acknowledged in his own memoirs. Instead, he should be recognized for his resistance to every means to push him from the academy or prevent him from graduating. He ultimately foiled those who wanted to see him fail, which was in itself a powerful form of resistance.

45. Marszalek, *Assault at West Point,* 253.

46. Albert E. Williams, *Black Warriors: Unique Units and Individuals* (Haverford, PA: Infinity Press, 2003), 22.

47. Marszalek, *Assault at West Point,* 255–65.

48. Michie, "Caste at West Point," 611.

49. Willard B. Gatewood Jr., "John Hanks Alexander of Arkansas: Second African American Graduate of West Point." *Arkansas Historical Quarterly* 41.2 (Summer 1982): 103–28.

TRYING TIMES ON
THE HUDSON

One Cadet's Witness to James Webster Smith's
Travails at Old West Point

Ronald G. Machoian

In the summer of 1870, only a few years after the Civil War's end, the United States was still absorbed by the conflict of Reconstruction politics. The role for Black men and women in American society was a question that pervaded nearly every discussion of the day's issues. Education, and higher education with it, was by no means exempt from this debate and often took center stage. As a national institution, the US Military Academy (USMA) at West Point, New York, was thrust to the forefront of this question when it became the scene of early partisan efforts to integrate the US Army officer corps. The record left by Cadet William Harding Carter, a young white man raised in slaveholding Tennessee, offers witness to this episode. His observations are the personal telling of West Point's struggle with its national role during a time that evoked complex reactions in both North and South, including among and even within the corps of cadets. Through Carter's eyes, Reconstruction's turmoil appears as an unwelcome interloper at a time when cadets were otherwise just trying to survive the daily rigors of cadet life. But welcome or not, race and the role of federalism in a country only recently torn by a civil war were discomforting issues that made their way to West Point that summer when the first Black American cadets reported.[1]

Attempts to racially integrate West Point set the stage for a longer struggle to integrate the US Army officer corps. When James Webster Smith became the first Black man matriculated to the academy, it created a firestorm that reflected the day's bigotry and defied the promise of radical Reconstruction.

Decades later, in 1910, the *New York Times* remembered the episode in succinct and plainly discouraged words that read more like an epitaph, "The case of Smith was a pathetic one."[2] Carter's personal record is one of many threads in which Smith's story is retold, showing how a white student culture, laden with decades of tradition, struggled to come to terms with the arrival of a young Black man among the corps of cadets. In the end, the corps failed to meet this test of humanity, and in the process, they collectively dishonored their institution and themselves.

On April 15, 1868, William Harding Carter received long-awaited word that he had been appointed a cadet by President Andrew Johnson. As a boy just a few years before, William had watched the Civil War unfold near his boyhood home just outside of Nashville. The son of one of the city's few Unionists, he watched as federal soldiers crowded the city, and he even served briefly as a dispatch rider for the Army of the Cumberland toward the end of the war. A hotel owned and operated by his father served at different times as a makeshift headquarters for Union commanders and their staff. After the war he remembered the soldiers with reverence and longed to someday become one of them. For him, a cadet appointment was the chance to become one of the army officers he had watched with such fascination.[3]

During the antebellum period West Point often served as a touchstone for critics who saw it as an affront to the meritocratic traditions embodied by the country's citizen-soldiers. The exodus of many academy men from US Army rolls to the war's Confederate cause had not helped its claim as a national repository of patriotism and service. The war's early years brought pointed assaults on the academy's national worth when West Pointers failed to produce immediate victories—a matter rectified to some extent later when graduates such as Ulysses S. Grant, George Gordon Meade, and William Tecumseh Sherman finally helped turn the war's tide in the Union's favor. But following the war, many skeptics, in celebration of the nation's volunteer tradition, still saw the school as a bastion of privilege, with only exaggerated claims to preparing men of any special military prowess. But this broader argument was not on Carter's mind as he prepared to take his place among the cadets.[4]

When Carter appeared with other candidates that June in 1868, he was arriving at a tradition-bound institution that was much the same as it had been for well over thirty years, since Sylvanus Thayer had transformed the academy while superintendent from 1817 to 1832. Thayer's legacy left the academy with demanding classroom standards and an austere barracks life.[5] Academics rested on the longstanding philosophy of "mental discipline," whereby cadets learned by reciting proofs, principles, or examples before the class section in a very precise manner under the instructor's critical eye—each cadet might be called upon to recite on any given day, leaving little opportunity to hide one's deficiencies. The objective was to "train the mind" to create an analytical approach toward problem-solving—whether in the classroom or later in the field. The curriculum was wholly prescribed for a cadet's entire four years; the sciences and mathematics accounted for the bulk of the coursework. Semiannual examinations were accomplished in each subject, and failure most often met with dismissal. The result was a course of study that demanded the "utmost energies both of students and instructors" in an environment that was described as "very severe and unrelenting."[6] Cadets had little time for recreation or socializing to find any respite from the classroom's daily grind. If there was a social awareness imbued at all, it was an unrelenting sense of continuity with the past and a feeling of shared perseverance.

Beyond activities such as horseback riding, barracks gambling, the occasional hosted dance with invited guests, or sometimes swimming in the Hudson River, one means of distraction was the fun cadets could extract from one another by various "games" or hazing. Hazing was by no means unique to the academy during that period, and the practice flourished at many colleges, especially in the East, as it had since the beginning of higher education in America. Given the well-defined student hierarchy at West Point, it of course featured prominently in cadet life. The practice of hazing or "deviling" welcomed each new class of freshmen, or "plebes," and this was perpetuated from year to year, a part of the collective experience. That West Point's cultural norms could create outright misery for one singled out as "different" was not a concern to Carter or seemingly anyone else at the time. For him, these traditions were an acceptable part of the place, and he looked forward to becoming a cadet despite the privations.[7]

Carter moved with his family to New York City shortly after the war to find greater opportunity and leave behind vanquished Nashville, now swollen with returning Confederate veterans and freedmen. The young man took advantage of the city's museums, libraries, and lyceums to continue his education where formal schooling left off. He began to record his thoughts on what he read and saw around him, a practice that he continued as a cadet. William, or "Willie," as his family knew him, did his best to prepare by finding out all he could about West Point and its traditions by reading anything he could lay his hands on about the place—paying special attention to Edward Boynton's *History of West Point* (1863), a work that celebrated the academy for its adherence to Thayer's vision.[8] But if West Point had changed little over the preceding decades, in 1868 the rest of the United States was struggling with a rapid pace of transformation across many elements of life. Many Americans, in both North and South, must have been anxious and discomforted by what the future might hold.[9]

The Fourteenth and Fifteenth constitutional amendments imposed federal authority to prohibit the states from denying suffrage based on race. But the promise held by each amendment was left largely unrealized in practice. The Fifteenth Amendment left open the legal tests of literacy, property, and education to gain the vote—avenues used in some former Confederate states to deny Black Americans equality despite federal intent. Southern Democrats worked to undermine the amendments at every turn while some whites turned to intimidation and violence to prevent Blacks from exercising the vote. In the North, attitudes toward racial equality were only marginally more progressive than in the former Confederacy. Although more opportunities were afforded Blacks in the North than elsewhere, white racial superiority remained a distinct boundary marking the nature and degree of those opportunities—a feeling that failed to change as the 1870s wore on. Even many northern whites grew tired of fighting the war's lingering battles, now defined primarily in racial terms. During Reconstruction, access to education remained one of the most evident chasms between social promise and its fulfilment. Segregated schools were the norm across even the North, with only the rare state-legislated exception. Freedmen considered access to education a doorway on their path to entering mainstream society on terms of equality, but white Americans were by no means prepared to share this opportunity on equal footing.[10]

American higher education reflected the era's racial divisions and remained overwhelmingly the arena of white men and a relatively few white women. Aside from a handful of traditionally liberal institutions, such as Ohio's Oberlin College, there were only rare instances of Blacks even receiving consideration for admission to white schools, let alone actually matriculating. The Morrill Land Grant Act of 1862 encouraged states to create land-grant colleges via the sale of federal lands, but even these schools remained segregated. Black Americans who had the preparation and means for higher education went to the era's emergent Black institutions, such as the Hampton Institute, Howard University, Atlanta University, or South Carolina A&M. In total, some two hundred different private and denominational Black colleges opened during the period. Not surprisingly, prior to the Civil War, only about two dozen free Blacks had graduated from white American colleges. During the thirty years that followed the war, it is estimated that only another two hundred did so—and seventy-five of those graduated from Oberlin. Harvard, already recognized as a leader in higher education, only graduated its first Black Bachelor of Arts student in 1870, and three decades later, its celebrated president, Charles W. Eliot, would still warn contemporaries that it was less than advisable to enroll more than a few Black students. It is not surprising that the cadet corps that Carter joined the summer of 1868 was no less white in its demography, even at an institution of distinct national purpose.[11]

Carter's first two years at West Point began his acculturation into the army, but his efforts in the classroom were marked by mediocrity. By the end of the spring semester of his second year, as a third-class (or "yearling") cadet, he was struggling to keep his chin above water in descriptive geometry. It was not entirely unexpected when he was finally found deficient in that subject during spring examinations. When the fall 1870 semester started a few months later, it was only through a merciful decision by the Academic Board (a powerful body consisting of the permanent professors of the several departments) that he remained a cadet at all. Carter had to repeat his third-class year, now as a member of the Class of 1873. The setback left him disappointed but more determined than ever to remain at the academy and graduate.

The year 1870 at West Point was important not only for Carter's personal survival as a cadet but also for the institution and the nation. President Grant and a Republican-dominated Congress pressed their agenda for Reconstruction while Democrats adamantly contested every step. On the Hudson River, for the first time in West Point's history, two Black men were nominated to the corps of cadets. That summer the two young men—James Webster Smith, son of a former slave, and Michael Howard, the son of a member of the Reconstruction Mississippi state legislature—made their way to West Point, where Carter and other cadets watched with great interest.[12] From this vantage point he witnessed the trials and tribulations of an attempt to effect social progress. As one contemporary editor put it with understated simplicity, "The application of the Fifteenth Amendment in the appointments to the Military Academy aroused much interest" from a variety of quarters. The episode became a microcosm of the day's broader issues—pitting federal racial reforms against deeply entrenched and vehemently protected American social norms. At that point efforts to integrate higher education grew almost entirely from federal activism, and so the fact that West Point, a national institution of some importance, became the site of similar efforts should have come as no surprise to anyone who was watching.[13]

A daunting and tense environment greeted Smith and Howard when they arrived at West Point in late June. Carter's own memory that "nothing in my time created more concern than the appointment of negro cadets" is a somewhat restrained description for the turmoil that would soon place cadets and faculty on a very public stage. As historian Scott Dillard summarized it, "The history of the Academy and its freedmen cadets is neither pleasant nor happy. Though it can be understood, it cannot be condoned." Academy leadership struggled with the challenge before them, mulling over with apprehension the line they should draw between enforcing a cadet's official treatment vice attempting to create a massive cultural shift. This threshold moment held untold measures of both promise and anxiety for West Point, the army, and American society.[14]

Carter and other cadets watched in shocked disbelief as the two young Black men took their place among the candidates for the Class of 1874. "We used to make jest over the mere idea of a negro cadet," Carter wrote home, "but the reality is to stern for a joke."[15] Early in June, the *New York Times* reported that a generalized feeling of "perplexity and anxiety" pervaded cadets as they considered the Black candidates' arrival.[16] Michael Howard, apparently an affable young man who might have been well-suited to the coming trials, never had the chance to make an attempt. He had only a year of formal schooling behind him and was eliminated by entrance examinations just four weeks after arrival. Better prepared scholastically, James Webster Smith, described by the *New National Era,* Frederick Douglass's own newspaper, as "nearly white" with features "more Mongolian than African," passed the exams and remained at the academy as its first matriculated Black cadet.[17] He owed his appointment to the support of David Clark, a wealthy liberal Republican who also had sponsored Smith's high school education in Connecticut, as well as to activism on his behalf by Gen. Oliver O. Howard, a one-armed Union war hero who was then chief of the Freedmen's Bureau. For his part, Carter looked on with agitation and regarded Smith as "the most unworthy fellow ever backed by misguided supporters."[18] The young man had enrolled in classes at Howard University just three weeks before receiving news that he had been nominated to West Point by South Carolina congressman Solomon L. Hoge, a Republican carpetbagger. It was clear to all that the appointment was a social-political gambit set on the national stage. By most accounts, Smith carried the academic training to succeed in the academy's demanding curriculum if given the chance, but beyond the cold and sometimes outright violent reception he received, his own reaction became another challenge to success, leading one West Point officer to describe him as "the personification of pulsive gloom."[19] His presence there was charged with anger and apprehension that loomed much larger than the simple appearance of a new cadet. Smith's own ability to cope would be put to an extreme test of personal endurance.

Only a few weeks after arriving at West Point, in a letter written to his sponsor, Smith reported horrendous mental abuse and maltreatment. Even more damagingly, these acts were alleged to have taken place with the full knowledge of staff and faculty who should have protected any cadet's person and position with their authority as army officers. Predictably, members of the Republican Congress were outraged when these allegations surfaced in public newspapers. Radicals Charles Sumner and Benjamin Butler even introduced legislation demanding a federal investigation, which, although passed, never brought any further action. Once the matter became public, Smith maintained that the contents of his letter had been "garbled" and didn't reflect his true feelings.[20] Satisfied by his own investigation that the charges lacked substance, Col. Thomas Pitcher, West Point's superintendent, nonetheless was stung by public criticism and realized that the issue of Smith's well-being was fraught with risk to the academy and likely his own professional reputation. Thinking preemptively, Pitcher convened an immediate board of inquiry to look into the matter.

The appointed board—consisting of two general officers and two field-grade officers—met and worked quickly but thoroughly, interviewing Smith as well as many other cadets, faculty, and staff at length. They found that although some level of mistreatment had indeed occurred, there was little to substantiate the charges as levied. Furthermore, the board held that the institution itself had treated Smith fairly and done its part to ensure his well-being. Additionally, the board recommended that Smith appear before a general court-martial for pressing his complaints outside the prescribed chain of command. Such military trials were commonplace during the period for even relatively minor cadet transgressions, and so this recommendation likely represented no special vindictiveness. Nothing more came of the episode after Secretary of War William Belknap reduced the remedy to a simple reprimand for all involved due to the convoluted circumstances, but it certainly forecast the nature of Smith's subsequent cadet career. The reprimand included the sons of at least two army general officers.[21]

There was an ironic prelude to that fall's events. As they prepared for recognition ceremonies, the Class of 1872 was sworn to "not interfere, molest, harass, & c. the new cadets" who thereafter would report as plebes—a point that Carter marked as a significant change, at least ostensibly, in the current state of things. The demanding system of training and discipline, based on class year, was already a source of anxiety to those in authority. Hazing had become more prevalent at West Point during the Civil War years, and despite leadership's efforts to curtail the practice, it survived and had its enthusiasts. This coerced oath indicated that hazing's persistence had become a challenge to good order in the corps, and Smith's arrival only exacerbated efforts to alleviate its terrors.[22]

In mid-August, about a month after the board of inquiry released its findings, Smith had gotten into a fight with a white cadet while still at summer encampment, employing a coconut water-dipper as a weapon to great effect, sending his foe, Cadet J. W. Wilson (also armed with a water dipper), to the surgeon's tent in the process.[23] He followed this episode two days later with a second, unrelated transgression by addressing an upper classman in what was reported as a "disrespectful" tone. Smith officially denied the latter allegation, but when three cadet witnesses corroborated his accuser, making a false statement was added to the list of charges. A court-martial was again the recommended forum, and this time, Secretary Belknap approved. Gen Oliver O. Howard, a public supporter of Smith's, presided over the court—drawing some onlookers' suspicions of a predetermined outcome. Nonetheless, Smith was found guilty of conduct prejudicial to good order and discipline for the assault but not guilty of making a false official statement. He was sentenced to walk punishment tours for six consecutive Saturdays for the offense, an outcome so light that it brought cries of indignation and disappointment from cadets and others. The judge advocate general, as reviewing authority, found the sentence so hollow as to make a mockery of the proceeding. He set it aside as "utterly insufficient" rather than offer the government's legal sanction, and Smith never marched a tour for the event. As the review moved through the War Department, cadets looked on with horror, leading Carter to write, "It has come to a disgraceful pass, when liars are kept in our midst. . . . But politics, that bane of an honest man who enters the public arena, has its influence upon the Sec'ty of War." Carter avoided acknowledging racial bigotry outright in his resentment for Smith and instead couched his reactions in terms of personal honor and "politics" as unwelcome influences and distractions.[24]

That fall Smith was placed at a mess table with several young men from the South—an arrangement that likely flowed from some purposeful mind. Despite his father's political leanings and his own service with the Union army, Carter, as a Tennessean, was assigned to this table as well. "I never knew who plotted the thing," Carter remembered years later, "but I do know we were better friends of the negro race than he was."[25] Some number of Smith's new "mess-mates" (perhaps including Carter) responded by requesting assignment to different tables, but their efforts were rebuffed by Brig. Gen. Emory Upton, new commandant of cadets at the school. Upton reminded the cadets that the academy's strict rule of order would prevail. The Union war hero and staunch abolitionist offered Smith encouraging words soon after his arrival at West Point. Upton also attempted to set the stage with a tersely worded speech that admonished the corps for personal insults against a brother cadet whose "only crime was color"—a point entirely missed by most of the audience. But if Upton thought then that speeches and talk would ensure fair treatment, he would be proven wrong.[26]

Smith's stay at the academy became a singular illustration to Carter, not only of the powerful changes at work in American society but also of the very pervasive reach of American politics in the nation's military affairs. It was this latter truth that proved a lifelong lesson for Carter, influencing much of his later public writing and efforts to spur military reform at the War Department. The lessons he learned at West Point were only the beginning of a long professional education founded in experience. But at the time, they only seemed a "bitter dose" pressed on him and his cadet counterparts by outsiders who simply did not understand or care to understand the rigors and privations of cadet life.[27]

Smith's treatment at the hands of upperclassmen ran a wide gamut—from being completely ignored to near-constant verbal abuse and provocation that made his daily life hardly bearable. White cadets responded with outrage that he did not suffer the tribulation quietly. "He is saucy and impudent, a confirmed liar," Carter reported to his diary, "it having been proved several times that he was a miserable liar." Cadets resented the fact that Smith might presume to seek the administration's shield, prompting Carter's complaint that he was "a tattler who runs to the Com'd't for the slightest thing which is done to him."[28] It must be remembered that all of this was suffered in an already-austere West Point culture that placed numerous challenges before cadets, each threatening their very existence at the academy if not successfully met. Cadets, at least white cadets, had one another to turn to for support and encouragement. But this sense of social support was completely denied to Smith and the other Black cadets who later followed him to West Point. Smith's circumstances were thus compounded to a degree that must have become nearly unbearable for a young man of his age, regardless of determination or preparation—leaving him, as one reporter observed, "more lonely than a solitary one in the desert."[29] In reaction, Smith stood up to his tormenters, playing directly into their purpose.

Relying on the reminiscences of Henry Ossian Flipper, the fourth young Black American to matriculate at West Point and Smith's roommate during the short time they overlapped there, it is difficult to avoid the impression that Smith's response to his ill-treatment likely exacerbated his situation. Unlike Flipper, who later found solace within himself, Smith fought back against his white counterparts—often with the expected outcome: "He demands much better treatment than his classmates," bemoaned Carter to his diary, "shades of chivalry and independence where art thou—then will no one rid us of a festering sore[?]"[30]

At this point, Carter's own reaction to Smith—like that of the entire corps of cadets—became one of absolute rejection. His diary entries on Smith's

presence reflect the period's pervasive social currents in both perspective
and expression. Carter had grown up with very positive feelings toward enslaved and later freed Blacks around him, becoming "inseparable" boyhood
friends with a young enslaved boy near his own age. In adulthood, he visited
his friend's family in Tennessee, learning that the man had named his daugh-
ter "Willie" after him.[31] But as a young cadet Carter failed to break from the
group-driven behaviors of his peers, not uncommon then or now. Their racial
attitudes were not out of the ordinary for the day, and in fact were very much
the norm across American social classes—North and South. As other contributors in this volume emphasize, many or even most white Americans did
not necessarily equate emancipation under the law with social equality. Four
years of violent war were only the start to a long and anguished fight for social
progress.

It is perhaps telling that Carter most often bemoaned Smith's existence in
terms of honor rather than debasing himself with overt racial arguments, even
if he used common racial epithets. Whether this was due to gnawing pangs
of moral conscience or genuine belief will remain unknown, but charges that
Smith allegedly lied to cover his reactions to the ceaseless harassment certainly
offered a convenient framework for the assertion. In nearly every diary entry
that contains reference to Smith as a "nigger," the term is surrounded by quotation marks—perhaps an indication that, even then, Carter was discomforted
by the day's hateful language and his own participation in Smith's torment.
Other period accounts recall that the entire corps almost invariably referred to
Smith as "nigger Jim"—not a surprising revelation in the context of the times
but a bitter reminder that cadets failed to see Smith as another cadet, a young
man who shared their trials and deserved no worse. The inhumanity of their
actions was lost to their racial animosity.[32] Cadets' seething complaints about
the admission of Black cadets to West Point were cloaked in a more complex
concern for the institution's moral integrity. Carter and his fellow white cadets
could rail on about the "liar" in their midst while ignoring the obvious ac-
companying commentary on the situation's injustice and their own behavior.
Referring to Smith's external supporters, Carter wrote, "Put them at the table
with the infernal liar and if they don't change their opinions in two minutes I
will relinquish mine forever."[33]

Carter's roommate was apparently somehow involved in one of the numerous altercations with Smith, reporting him twice for lying. As Smith endured
each process, other cadets watched with measures of both satisfaction and
disgust. "The court martial for the 'negro cadet' has just adjourned," Carter
reveled. "He was tried for lying, and proven guilty." In a letter home, written

in juxtaposition or perhaps as a sort of justification, he noted that three white cadets had recently been "run off" by the first class (or senior cadets) for lying. This episode, which took place in January 1871, was one which provided yet another distraction for West Point that winter and drew a great deal of unwanted external interest—events that, for Carter, underlined the fact that military service carried with it a context of public interest.[34]

A congressional committee appointed to investigate this episode identified three young men as ringleaders of the vigilante action. It seems that three plebes made false official statements in their habitual act of disappearing to have a few drinks at a local tavern. In this instance, two of the three departed the academy grounds while the third covered for his thirsty classmates by reporting them "present and accounted for" to that night's cadet guard. Rather than await formal due process to take its tiresome course, nearly the entire first class—totaling thirty-eight cadets—took matters into their own hands to rid the corps of the three whose honor was suspect.[35]

These events were quickly reported to Colonel Pitcher, superintendent at the time, as well as to the press. The vigilante upperclassmen claimed a higher moral purpose for their actions, and the rest of the corps, with Carter among them, lined up in overwhelming support. In the post–Civil War decades, West Point's cadet honor code was still developing and lacked codification beyond leadership's tacit approval in both word and act. Cadets were, however, very much aware at that point of the sober expectation that they act with an officer's gentlemanly honor and deportment—expressed in various traditions, symbols, and practices for almost a century before becoming an actual code.[36] Already cynical about the public's treatment of events at West Point, Carter complained to his family about this latest episode in terms that reflect the prevailing view among his peers: "I suppose you have seen a great deal in the papers concerning various matters at West Point of late, and I might add with perfect fairness, that you have not seen much truth in any of it. I will give you a few particulars in which the papers are faulty. They state that those cadets who were run off were so treated because they got drunk; this is very far away from the true state of the case." He then went on to explain the situation in terms very similar to those he had used when discussing Smith's own transgressions. A cadet's honor, though lacking definition at the time, was, at the very least, easily invoked as a venerable shield whenever the hour demanded: "They were kicked out because they were a set of liars. There have been more men reported for making false statements in the present plebe class than in all the others I suppose for ten years." That vigilantism sapped the remedy of its claimed sanctified purpose was lost on Carter and likely the others involved.

"Looking at the case in this light," he wrote, "it was decided to take the matter into our own hands, and save the trouble of a court martial."[37]

But members of Congress were not as certain as Carter and his fellow cadets. A committee was formed to accommodate public interest, and its report was stridently critical of the academy's leadership for its failure to enforce order while seeming to indulge mob law on the bluffs overlooking the Hudson River. In the end, the three cadets were reinstated, and little happened to the involved first classmen, although rumors of mass expulsion must have provided some anxiety for them only a few months before graduation. This case, and the attention it brought from eastern reporters, was only a sideshow to the travails of James Webster Smith, but it reflects the spirit of a kind of vigilante determination that existed within the corps at the time. Writing again of the Smith case, Carter noted that if the sentence was again met with indulgence by the secretary, cadets might similarly take matters into their own hands, "There is a silent determination gaining hold of the Corps every day," he wrote, "if such an occurrence verifies itself, it behooves us to 'spirit him off to the mountains' also. The firm discipline here makes men bear a great deal, but there is a point when forbearance is no longer a virtue."[38] The irony that "disciplined" cadets would act extralegally was missed by Carter.

Some related the vigilante incident to Smith's own circumstance, charging that cadets would cynically normalize Smith's treatment in terms of the punishment meted out to the white "liars" who were summarily run off. Whether he consciously rationalized the circumstances by way of comparison is unknown, but certainly Carter viewed both instances as the cadet corps' reasonable reactions to external meddling and institutional impotence. "Every country Congressman, Politician and Editor seems to think [it] his duty to his constituents, God, his country and himself, to give some advice concerning the management of West Point," he groused, "while in fact [they] have no more idea of the workings of the academy than a hog has of music." Carter's recorded thoughts at this point, if nothing else, most certainly reveal a young man plainly convinced that cadets knew best about cadet doings, and others should resist the urge to interfere.[39]

Carter protested that the concept of honor was rendered farcical by recent lack of enforcement and that the still-evolving honor code's execution should reside within the corps itself. Testimonials to the investigating committee reflected widespread dissatisfaction with the lack of punishment for transgressors, and likely Smith's continued presence among them was foremost on their minds. Observers reported this perspective prevailed among the entire corps, with common assertions that Smith's transgressions "sullied" his honor and thus so too the honor and dignity of all. Carter's characterization of Smith's conduct in terms of breached integrity reflected a common refrain among cadets at the time. Though often punctuated with clear racial overtones—"an impudent negro liar protected by the authorities is too much for anybody to stand"—their disdain was professed in the language of West Point's claimed emphasis on forthright deportment as part of its developing creed. "Put them [external observers] at the table as I am, with the infernal liar," Carter grumbled, "and if they don't change their opinions in two minutes I will relinquish mine forever." Congressional oversight and especially Smith's external support among Radical Republicans was a sharp slap to cadets who believed they alone should act as West Point's institutional gatekeepers, even at an academy sustained by public charter for national purpose. Carter hoped that Congress would simply "keep its soiled fingers out of [academy] affairs," pleading that control should remain with the War Department, and not be subjected to the "political spoils" of republican government.[40] The fact that cadets' reactions to Smith were dishonorable, rejecting him outright for his race while making his life miserable in the process, was lost amid a sense of misplaced collective morality, rationalized by the period's racial norms.

In December, after an incident that started when a white cadet kicked or stepped on Smith's feet while marching in formation, Smith was brought up on charges that included fighting and inattention in the ranks. After denying these allegations to the commandant of cadets in the face of much testimony otherwise, Smith was again charged with making false official statements. Emory Upton dutifully enforced cadet regulations, and Smith soon found himself before another court-martial board. In January 1871, after serving as his own defense counsel, Smith was convicted of "conduct unbecoming a gentleman" as well as lying.[41] Carter, like everyone else at the academy, watched these events critically with vested interest. In a letter written to his mother, he bristled that a Black cadet remained among them, normalizing his thoughts again in terms of disgust for politicized external pressures. "We are waiting anxiously for the court-martial orders of the 'American gentleman of African descent.' I believe his sentence was dismissal, and it is to be hoped that the late developments of the interference of Congressmen, will prevent his sentence from being commuted."[42] He wrote with particular disdain for David Clark, Smith's benefactor, indignant that Clark would advocate true equality for Smith within the corps:

> I think that the wretch who proposed in Congress that "Mr. Clark be allowed to testify that Cadet Smith was incapable of falsehood" ought to

be hung, if not really at least killed politically. It is indeed strange to me how decent men can stand up for Clark and his protégé after his statement in which he says words to the effect that he considered that the negro should be accepted on terms of social equality by Cadets, for said he "I allowed him to sit at [the] table with coachmen and gardener, both intelligent American citizens." He thus has the extreme kindness to put us [the cadet corps] on an equality with his groom and gardener; nice comparison for men expected to lead the army of the Great American Republic noted particularly for . . . the rascality of its Politicians.[43]

The board indeed sentenced Smith to dismissal, a predictable outcome for the compounded charges. But the sentence was suspended until the end of the academic year while those in Washington, including President Grant himself, discussed the appropriate—or, more likely, the least politically damaging—course of action.

Meanwhile, Smith's daily life remained one of complete ostracism and isolation, allowing him to avoid further confrontation but depriving him of any society whatsoever. In mid-March, Carter answered his mother's inquiry about Smith's situation by stating simply, "He still survives, but no one ever takes any notice of him because we are so disgusted with the authorities at Washington." Once again, while indignant at the politics that kept Smith among them, Carter avoided addressing head-on the unstated but powerful racial forces at play. "The 'anomaly' as we call him at the table when speaking of him (no one ever speaks to him), has been tried so often and the sentences so smothered down in Washington that we have given up all hopes." It is ironic, but not surprising, that in this same letter Carter follows the above comments with a glowing summary of that Sunday's sermon—apparently not finding any conflict between cadets' cruel daily treatment of Smith and the upright dictums of moral virtue; an irony certainly not limited to Carter and his fellow cadets in any decade. In another instance, young Carter even prayed to God for Smith's dismissal, an absurdly purposeful request for Divine intervention. In the meantime, cadets, faculty, and staff waited for the administration's decision while preparing for spring examinations. "The sentence of the last court-martial was sent to the War Department several months ago, and should, in the natural course of pigeon holes, have found its way out in two weeks," Carter wrote, reflecting the cynicism and impatience shared in the cadet barracks, "hence we can only conclude that there's foul play," in reference to the politics surrounding Smith's circumstances.[44]

In April, Smith was charged with yet another offense after applying "an ungentlemanly epithet to another cadet," and then responding in a "highly indecent and vulgar manner" when verbally reprimanded by the section marcher. "I cannot be more explicit for the affair is a disgrace to the Corps," spluttered Carter, "if a scoundrel of his stamp can be said to disgrace those who are forced to be his 'companions' in arms."[45] This charge was apparently never brought to trial, perhaps because Smith was already embroiled in such serious trouble.[46]

In June 1871, sensitive to the larger ramifications surrounding Smith's continued existence at West Point and coming under increasing political pressure, the Grant administration finally acted and reduced Smith's sentence of dismissal to the more palatable punishment of repeating his plebe year.[47] Carter took the opportunity to pounce once again on what he considered the galling nuisance of the nation's larger social-political interest in academy affairs: "The negro's orders came through this afternoon. He was sentenced to be dismissed, but it was commuted to being turned back. I was surprised at the frankness of the Sec'ty of War. He said as Smith 'the negro' was the representative of the policy of the administration and it would better the ends of this negro-loving policy to forgive him his third offense for lying, than to send him away. Justice where art thou?" Such criticism misses completely the fact that West Point, by the very structure of its political admissions and role as a national university, was an extension of American society and thus plainly beholden to its political authority and shifting social course. But Carter and his fellow cadets failed to see beyond the cadet barracks' dominant subculture and its blinding racist lens.[48]

At the end of the academic year, Smith ranked thirtieth among his class of fifty-five in mathematics and twentieth in French—fair grades considering the negatively charged atmosphere which no doubt distracted him at every crack of a book. The repeat of his plebe-year academics saw great improvement, even if in subjects in which he had already once prevailed. It was during the 1871–72 school year that Smith was joined by another young Black American man at the academy, Henry Alonzo Napier. Napier's appearance afforded some sense of shared adversity and society, but he lasted only a year there. Of Napier's arrival, Carter took only very brief note in his diary, stating that "the 'nigger' is amongst the sixty who got through" in reference to that summer's entrance examinations and matriculation.[49] In 1872, another Black cadet, Thomas Van Rensselaer Gibbs, entered West Point but only remained through the fall semester before departing for failure in mathematics.[50] But just a few months later, in May 1873, two more Black men arrived, John W. Williams, who was dismissed for academics after only a single semester, and Henry Ossian Flipper, who would become West Point's first Black American graduate and part of the country's story of racial perseverance. Flipper, Williams, and Smith

roomed together, but even Flipper noted Smith's apparent inability to accommodate the situation—battling each hateful slight instead of pressing onward
sustained by their larger purpose. Smith had written to Flipper upon his arrival
there, urging him to "avoid certain consequences" that would accompany "forward conduct"—advice that Flipper took to indicate some element of Smith's
private regret for his own failure in that same vein, at least during his first
year as a cadet. Given almost constant provocations from the very beginning,
Smith appeared unable to ignore his tormenters, an approach that only compounded the harassment he endured.[51]

Despite Cadet Flipper's presence and the relative calm of his ensuing years
at West Point, Smith did not make it to graduation. The script for the next
two years was one of continued ostracism accompanied by steadily deterio-
rating classroom performance and class standing. In the spring of 1874, he
was "found" deficient academically after prolonged problems in natural and
experimental philosophy. The automatic outcome for this failure was disen-
rollment, unless the Academic Board recommended otherwise—something
the inveterate professors rarely entertained for one already turned back previously. The decision was upheld at the War Department, and in late June, James
Webster Smith left West Point four years after his arrival and returned to his
native South Carolina.[52] Smith, noted the *Hartford Daily Courant,* "got his
fill of West Point and its glories, which to him have always come in the shape
of tribulations and trials." In a series of editorials, the *New National Era* lamented the ending to Smith's cadet career as part of the larger social travesty
of Reconstruction's unfilled promise, damning his treatment and expulsion
as the product of "a miserable spirit inconsistent with manly bravery, and an
exhibition of moral cowardice." With ill-concealed satisfaction, the editors of
the *Army and Navy Journal* published an epitaph shared by many predisposed
toward the young man's failure, "The disease of 'civil rights,' as applied to the
selection of cadets, left to run its course, has wrought its own cure, and the
law of the 'the survival of the fittest,' has brought about a most desirable result
without doing violence, except it be philanthropic desire."[53]

The next year, Smith accepted a position at South Carolina Agricultural
and Mechanics Institute at Orangeburg, serving as both its commandant of cadets and an instructor of mathematics. He wrote scathing public letters about
his time at West Point and steadfastly refuted suggestions that his problems
grew from his own haughty demeanor. He damned the administration and
the secretary of war for not having intervened to reinstate him a second time,
and attacked faculty, charging them as complicit with the cadets' abuse. These
assertions, however accurate, failed to gain any official support or even much
sympathy among the white reading public. Smith's troubled story ended sadly

just two years later, on November 30, 1876, when he died prematurely from tuberculosis.[54]

Carter had graduated in 1873, a year before Smith's dismissal, and taken his place in the frontier army as a lieutenant in the US Eighth Infantry, then posted at Fort D. A. Russell in the Wyoming Territory. He was not present to witness Smith's departure from West Point or to record his concluding thoughts on the episode. Carter's diary entries and surviving letters written in 1872 and 1873 contain no further references on the subject; perhaps because Smith's existence in the barracks had by then become less of a novel insult and the topic had grown stale. But the deepest travesty in this narrative is not so much that a single Black cadet was subjected to attacks, social isolation, and pressures that proved beyond his personal endurance. If so, Smith's story would be only one of the uncountable humiliating footnotes to the day's annals of racial hatred and injustice. Instead, Smith's cadet career is part of a larger story, one in which West Point, one of our nation's most revered institutions, failed to meet what was perhaps one of its most daunting challenges of the era.

In Smith's case, the record stands on its own that academy leadership did little, or at least not enough, to exercise whatever authority and levers were at its disposal to shape the student culture in a manner that might have overcome powerfully ingrained racial attitudes. At first consideration, this failure is easy to excuse as an unhappy byproduct of the times. After all, at some level, the corps of cadets was a microcosm of the society from which it was drawn, and America's racial bifurcation is a matter of historical record. But while difficult to prosecute from the distance of more than a century's hindsight, it is inescapable that becoming a West Point cadet even then rested on the very premise that training and acculturation would instill institutional values and demand specific behaviors. In many ways, these values and behaviors often departed from society's norms to embrace those of the institution—as true then as it is now in officer training programs. Was not the period's racial hatred a norm that might have been overcome or at least mitigated with stronger, more earnest leadership?

Emory Upton, commandant of cadets, was an upright abolitionist who rarely shied from any sort of challenge, whether on or off the battlefield. Upton himself had been ostracized by others as a cadet due to his unashamedly abolitionist ideals. Had he been placed at West Point as commandant solely for the purpose of championing cultural shift to effect racial integration, there could have been no better choice in the army.[55] Upton undoubtedly was aware of Smith's daily misery, given the number of complaints that made their way to his desk. At times he appears to have made real attempts to mitigate the cadets' rejection of Smith, even arresting white cadets who were alleged to have singled him out for ill treatment. It is claimed that Upton offered Smith complete justice and reassured him that "you shall not be persecuted into re-signing. I am your friend. . . . Come to me and you shall have justice."[56] But as events transpired, Upton's complete faith in West Point and its cadets was deceptively optimistic and may have prevented him from seeing the gravity of the circumstances. Simple enforcement under the regulations proved not nearly enough to offer Smith a fair and competitive field.[57] While comman-dant, Upton took great personal pains to "influence cadets' behavior through religion," wrote David Fitzpatrick, his biographer. Might this same missionary zeal have been applied to mollify cadets' treatment of Smith and demand the same diligent oversight by staff and faculty on his behalf?[58]

That Upton as commandant might have felt shackled by lack of public support for more ambitious efforts is somewhat more believable, but it is still a thin explanation owing to common knowledge that the administration was under immense partisan pressure to see Smith's success. President Grant's per-sonal support for Smith may have been restrained early on by the fact that his own son, Fred, a member of the Class of 1871, was a cadet and may have been one of Smith's alleged tormenters.[59] But that interesting element notwith-standing, these are not just critical ramblings levied from historical distance—the seeming absence of vigor shown by those in authority was also noticed at the time. Many onlookers were outraged that West Point's leaders did little to tip the balance of justice toward some semblance of fairness. A *New York Times* correspondent alleged that Smith lacked support among the academy's officers and that faculty attitudes toward racial integration were very much the same as those found among cadets—looking away in complicit silence. The *New Na-tional Era* was outraged that Smith had not found more support among the staff and faculty: "The officers having control of the Military Academy seem to have absolute power," yet "there seems to be none to whom the colored citizen can appeal for justice in an institution that is the common property of all citizens."[60] While they dutifully enforced regulations, and some few—at least Upton and perhaps also Superintendent Thomas Pitcher—may privately have hoped for Smith's success and attempted due process, it is clear that the academy's leaders failed to demand an atmosphere of greater tolerance among cadets and perhaps faculty and staff as well. This can only be seen as a glaring dereliction in their responsibilities even given the context of the times—an argument taken up further by other contributors to this volume.[61]

In Carter's testimony, the individual, although a young man susceptible to the social pressures of the cadet barracks, is seen as an exemplar of unful-filled hopes. Carter was well read and insightful, undoubtedly exposed to the reasoning of abolitionist morality as well as witness to his father's courageous Unionist stance in Confederate Nashville. That he, like his peers, would wrap intolerance in self-righteous disgust at external "meddling" in the academy's affairs is sad irony. For them, Black Americans remained decidedly less than equal, despite their freedom. Resentment for Smith was held upright behind contempt for his alleged transgressions of the still ill-defined honor code. Cadets were indignant that outsiders would dare keep the dishonorable pariah among them, even as they sullied their institution's professed tenets with acts and words of bigotry and hatred for one of their own. Carter rationalized his feelings against Smith in terms of the academy's greater good, full of resentment that it was made a battlefield in the partisan quest for racial equality.

Later, as an army officer, in several instances (most notably while posted in the Arizona Territory and later in the Philippines on Samar) he championed the rights of marginalized peoples subjected to degrading injustice—and did so vocally at risk to his military career.[62] His experiences in the army matured his outlook and gave rise to a more thoughtful, confidently asserted sense of fair play. After serving for more than two decades on the frontier, Carter's posting to the adjutant general's office in 1897 placed him near the nation's senior decision-makers, where he found a receptive audience for his ideas on the developing military profession. He played a leading role in the series of progressive acts known as the Root Reforms, becoming an emphatic voice in public arguments for a modernized military organization. Ironically, the presence of politics in the nation's military affairs accompanied nearly his every endeavor as a senior officer, and he often referred to his West Point years as a baptism in this frustrating reality—one which he became very adept at accommodating during his time near Secretary of War Elihu Root.

Carter died in 1925 after a long and successful career as a soldier. His legacy as a reformer reflects the army's own transition to twentieth-century modernity. But his time near James Webster Smith while a cadet remains a regrettable footnote, just as it does for the institution itself. Perhaps any critical expectation that West Point and its cadets might have overcome the country's prevalent social attitudes to lead racial integration forward is too much to expect. That certainly is the vein of its contemporary defense. After all, the integration of American higher education took place only in the barest fits and starts at even the most liberal of colleges until well into the twentieth century. Surely there is an element worthy of admiration in West Point's mere attempt to integrate at that early date, despite its thorough lack of success. Only three Black men were graduated and commissioned from the academy during the two decades following Smith's dismissal, and it would be many more years before the academy truly integrated—not the outcome that Smith's advocates had in mind that summer in 1870.[63]

At its conclusion, then, Smith's story is one of lost opportunity, a failure paralleled by the story of Reconstruction more generally. Although slavery was eradicated by Union battlefield victories, race remained a pernicious barrier that relegated Black Americans to unequal footing in both public and private spheres. Higher education was no different in this regard, and even West Point, an institution of national purpose, reflected that cruel travesty. Smith's experiences as a cadet, in a very personal manner, met Reconstruction's challenges head-on and found racism simply too great a hurdle to overcome, not only for him and his white counterparts, but for West Point, the army, and the country as well.

Notes

1. There is a large body of scholarly literature covering the immense subject of post–Civil War Reconstruction. As excellent starting points, see Eric Foner's *Reconstruction: America's Unfinished Revolution, 1863–1877* (New York: Harper and Row, 1988), and also the abridged edition, *A Short History of Reconstruction, 1863–1877* (New York: Harper and Row, 1990). A more recent collection of essays by Eric Foner and Joshua Brown relate Reconstruction's social-political themes to contemporary events and practices: *Forever Free: The Story of Emancipation and Proclamation* (New York: Knopf, 2005). More specific to executive issues of Reconstruction politics is Brooks D. Simpson's *The Reconstruction Presidents* (Lawrence: University Press of Kansas, 1998). Leslie H. Fishel Jr. addresses the postwar culture of prejudice that quickly enveloped the United States in "The African-American Experience," in *The Gilded Age: Essays on the Origins of Modern America,* ed. Charles W. Calhoun, 137–61 (Wilmington, DE: Scholarly Resources, 1996). Also on postwar America and race, see John Hope Franklin and Alfred A. Moss Jr., *From Slavery to Freedom: A History of African Americans,* 8th ed. (New York: Knopf, 2000).

2. "West Point May Get a Colored Cadet," *New York Times,* January 24, 1910.

3. William H. Carter, Memoirs, 12–26. Carter's memoirs remain unpublished. The associated pagination noted throughout this article refers to that found in the original memoir typescript. The Carter memoirs, diaries, and letters referenced herein are part of a private collection gifted to the archives of the US Army's Heritage and Education Center (AHEC), Carlisle, PA. This collection was used by this author while still in the private domain. The unedited memoirs, present in the collection gifted to AHEC, cover nearly Carter's entire life, from youth through his second and final retirement in 1918. Carter died in 1925. The diaries begin in 1868, as Carter prepared for West Point, and continue through his graduation, although without consistent entries. Carter's family owned and operated Nashville's St. Cloud Hotel,

a place that served as headquarters for Union general Don Carlos Buell and was frequented throughout the war by Union officers. See Ronald Machoian, *William Harding Carter and the American Army: A Soldier's Story* (Norman: University of Oklahoma Press, 2006), 13.

4. On the controversy surrounding wartime West Point, see T. Harry Williams, "The Attack upon West Point during the Civil War," *Mississippi Valley Historical Review* 25.4 (1939): 491–504; Alan Aimone and Barbara Aimone, "Much to Sadden and Little to Cheer: The Civil War Years at West Point," *Blue and Gray Magazine* 9.2 (December 1991); and Lori Lisowski, "The Future of West Point: Senate Debates on the Military Academy during the Civil War," *Civil War History* 34.1 (1988): 5–21. To better understand USMA's institutional perspective of external pressures and the day's issues, see George L. Andrews, "The Military Academy and Its Requirements," *Journal of the Military Service Institute of the United States* 4 (1883): 112–46. There are numerous entries of this sort found in post–Civil War periodicals such as *Army and Navy Journal* and *JMSIUS* as well as letters written to major daily papers arguing West Point's purposeful role and its value to American society. For instructive examples, see "Military Education," *Army and Navy Journal* 8.28 (March 3, 1866); and "West Point Training," *Army and Navy Journal* 8.29 (March 10, 1866). James L. Morrison discusses contemporary criticism of West Point in "The Struggle between Sectionalism and Nationalism at Ante-Bellum West Point, 1830–1861," *Civil War History* 19.2 (1973): 138–48; as does Walter Scott Dillard in "The United States Military Academy, 1865–1900: The Uncertain Years" (PhD diss., University of Washington, 1972), 98–129. Dillard devotes a chapter of his dissertation to "The Black Cadets' Dilemma, Agony, and Failure" (183–225) and presents what is likely the best critical narrative of these events to date. William S. McFeely, *Grant: A Biography* (New York: W.W. Norton, 1981), examines Smith's introduction at USMA from the administration's perspective; and John F. Marszalek details the experiences and eventual court-martial of Cadet Johnson Chesnut Whittaker, a young Black man who entered West Point in 1876, in *Court Martial: A Black Man in America* (New York: Charles Scribner's Sons, 1972). Several published narrative works are also useful—among them are James L. Morrison's *The Best School in the World: West Point, the Pre-Civil War Years, 1833–1866* (Kent, OH: Kent State University Press, 1986); Stephen Ambrose, *Duty, Honor, Country: A History of West Point* (Baltimore: Johns Hopkins University Press, 1966); Thomas Fleming, *West Point: The Men and Times of the United States Military Academy* (New York: William Morrow, 1969); and George Pappas, *To the Point: The United States Military Academy, 1802–1902* (Westport, CT: Praeger, 1993).

5. On the Sylvanus Thayer years at West Point, in addition to the resources cited in note 4 above, see especially Edgar Denton, "The Formative Years of the United States Military Academy, 1775–1833" (PhD diss., Syracuse University, 1964). Upon his arrival as superintendent in 1817, Thayer began an era of reform at early West Point that created a foundational academic and military culture that largely survived the nineteenth century and, in some respects, remains evident even today.

6. For a contemporary description of USMA's classroom practices, see Charles W. Larned, "The Genius of West Point," in *The Centennial of the United States Military Academy at West Point, New York: 1802–1902* (Government Printing Office, 1904; reprint, New York: Greenwood Press, 1969): 467–507, quote p. 477.

7. Hazing was a central feature of nineteenth-century college life—at USMA, see Dillard, "Military Academy," 292–311; Pappas, *To the Point*, 343–45, 355–56, 412–15; and for a contemporary address on the practice at civilian institutions, "College Hazing," *Scribner's Monthly* 17.3 (January 1879): 331–33. This last article notes that hazing had begun to subside during the 1870s but was still prevalent at many eastern colleges, though seen only rarely in the newer western universities. Helen Lefkowitz Horowitz narrates the historical development of student culture at the American college and university in *Campus Life: Undergraduate Cultures from the End of the Eighteenth Century to the Present* (Chicago: University of Chicago Press, 1987), and posits class year–based hazing and other student norms as part of the campus acculturation process during the nineteenth and early twentieth centuries.

8. Edward C. Boynton, *History of West Point, and Its Military Importance during the American Revolution: And the Origin and Progress of the United States Military Academy* (New York: D. Van Nostrand, 1863; reprint, Freeport, NY: Books for Libraries Press, 1970).

9. Carter's diaries during the period reflect a developing enthusiasm for the day's literature, art, theater and even philosophy. In 1867–68 he made great use of local libraries and later kept a list of books and journals he had read, one of his claimed favorites being the *Phrenological Journal*. Carter at times reflected on contemporary events and the recent war's destruction. (For example, see diary entries of November 19 and 27, 1867, and January 1 and 19 and February 13, 1868.). A copybook kept with Carter's diaries contains a short "autobiographical" essay written on April 25, 1869. On the despair and social divide that gripped postwar Nashville, see Peter Maslowski, *Treason Must Be Made Odious: Military Occupation and Wartime Reconstruction in Nashville, Tennessee, 1862–1865* (Millwood, NY: KTO Press, 1978). Carter's immediate family was broken by the war: an older sister was married to a Confederate officer. She never spoke again to or communicated in any manner with the rest of the family. See Carter, Memoirs, 13.

10. Foner and Brown, *Forever Free*, 162, 192. For detailed chronology and discussion of state actions pursuant to accessible education, see William Preston Vaughn, *Schools for All: The Blacks and Public Education in the South, 1865–1877* (Lexington: University Press of Kentucky, 1974). Most recently, Millington William Bergeson-Lockwood explored race as an influence on post–Civil War politics in "Not as Supplicants, but as Citizens: Race, Party, and African American Politics, in Boston, Massachusetts, 1864–1903 (PhD diss., University of Michigan, 2011). Bergeson-Lockwood supports Eric Foner's conclusion that racial attitudes toward Blacks in the North—and thus also opportunities—were not much different than elsewhere during Reconstruction; see Foner's chapter "The Reconstruction of the North" in *Reconstruction*, 460–511.

On the subject of failed Black leadership in a racially segregated postwar North, see Leslie H. Fishel Jr., "Repercussions of Reconstruction: The Northern Negro, 1870–1883," *Civil War History* 14 (December 1968): 325–45.

11. On freedmen's pronounced desire for education following emancipation, see Foner and Brown, *Forever Free*, 88–89, 162–63. More generally, on African American postsecondary education in the United States, see Frederick Rudolph's classic historical work, *The American College and University* (New York: Knopf, 1962; reprint, Athens: University of Georgia Press, 1990), 488–89; Arthur M. Cohen, *The Shaping of American Higher Education: Emergence and Growth of the Contemporary System* (San Francisco: Jossey-Bass Publishers, 1998), 110–11; Alan Pifer, *The Higher Education of Blacks in the United States,* reprint of Hoernle Memorial Lecture for 1973 (New York: Carnegie Foundation, 1973); John R. Thelin, *A History of American Higher Education* (Baltimore: Johns Hopkins University Press, 2004); Christopher Lucas, *American Higher Education: A History* (New York: Palgrave-MacMillan, 2006); Vaughn, *Schools for All,* 103–18; and also Caroline Sotello and Ed Turner, eds., *Racial and Ethnic Diversity in Higher Education* (Needham Heights, MA: Simon and Schuster, 1996). This latter resource is a collection of seminal essays relating to the subject. Also see Wesley Brown, "Eleven Men of West Point," *Negro History Bulletin* 19.7 (April 1956): 148. The estimate of two hundred Black graduates during the years 1865–1895 and that seventy-five of these were from Oberlin is found in Pifer, *Higher Education of Blacks,* 13. Thelin details the evolving role of federal agency via the Morrill Act in *American Higher Education,* 75–81; Charles Eliot's public assertions on race are noted in Thelin's same book, 173. Eliot's evolving position on race in higher education is examined by Jennings L. Wagoner Jr. in "The American Compromise: Charles W. Eliot, Black Education, and the New South," in *Education and the Rise of the New South,* eds. Ronald K. Goodenow and Arthur O. White, 27–41 (Boston: G. K. Hall, 1981). On Harvard College's first Black student to graduate with a bachelor's degree, see Katherine Reynolds Chaddock, *Uncompromising Activist: Richard Greener, First Black Graduate of Harvard College* (Baltimore: Johns Hopkins University Press, 2017). Three Black men had graduated from Harvard's professional schools a year earlier, in 1869: Edwin C. J. Howard (medicine), George L. Ruffin (law), and Robert T. Freeman (dentistry). Harvard's story with race is told through a collection of edited primary materials in Werner Sollors, Caldwell Titcomb, Thomas Underwood, and Randall Kennedy, eds., *Blacks at Harvard: A Documentary History of African American Experience at Harvard and Radcliffe* (New York: NYU Press, 1993). Harvard enrolled three Black men as early as 1850, but they did not remain long—see *Blacks at Harvard,* 2.

12. Adam Domby details Smith's background as well as offering more information on Samuel Hoge, the congressman who appointed Smith, in his essay in this volume, "A Nursery of Treason Remade?"

13. Editorial comment, *Army and Navy Journal* 8.42 (June 4, 1870). For discussion of integration in colleges as the result of broader federal efforts to influence reform, see Pifer, *Higher Education of Blacks;* and Vaughan, *Schools for All.*

14. Carter, Memoirs, 43; Dillard, "Military Academy," 188. The *New York Times* discussed a proper policy or approach to Smith's presence by USMA's senior leadership in a short editorial posted on June 10, 1870: see "West Point: The Colored Applicants for Cadetship." National newspapers, including the *Times,* its competitor the *New York Tribune,* and the *Hartford Daily Courant* (due to David Clark's interest) began covering Smith's trials at West Point on a regular basis. Archival holdings of both the *New York Times* and the *Harford Daily Courant* were used to help frame this study.

15. Carter diary, November 10, 1870.

16. "West Point," *New York Times,* June 10, 1870.

17. *New National Era,* "West Point," January 12, 1871.

18. Carter, Memoirs, 43.

19. Capt. Rufus King's racially punctuated descriptions of both Smith and Howard are found in Fleming, *West Point,* 214. On Smith's background, see William P. Vaughan, "West Point and the First Negro Cadet," *Journal of the American Military Institute* (October 1971): 99–102. The *Hartford Daily Courant* detailed Smith's high school preparation in "The Colored Cadet Again," June 20, 1870.

20. "The Colored Cadet," *New York Times,* July 15, 1870.

21. See Vaughn, "First Negro Cadet"; Pappas, *To the Point,* 376–77; and Dillard, "Military Academy," 196. The board of inquiry, seated at Colonel Pitcher's request, was chaired by Civil War hero James Harrison Wilson; Gen. H. L. Abbott and Maj. T. F. Rodenbough served as members at large; and Maj. Thomas F. Barr acted as the judge advocate representative.

22. Carter diary, February 25, 1869; on the progression of hazing at West Point, see Pappas, *To the Point,* 343–46; and Ambrose, *Duty, Honor, Country,* 222–31. Carter remembered the hazing rituals at West Point in Memoirs, 34.

23. "From South Carolina: Letter from Cadet Smith," *New National Era,* August 6, 1874. In this letter, written after his dismissal from West Point, Smith recalls the verbal and physical assaults of the 1870 summer encampment.

24. Carter diary, November 10, 1870. On the court, see Vaughan, "First Negro Cadet"; and Dillard, "Military Academy," 196–97. For representative news accounts and editorials on the proceedings, see the *New York Times,* October 21, 26, and 28, and November 17, 1870; and *New National Era,* November 17, 1870.

25. Carter, Memoirs, 44.

26. "The Colored Cadet," *New National Era,* July 21, 1870. The *New National Era* was published by Frederick Douglass from 1870–74 as a progressive organ that gave voice to freedman's issues and interests across a broad spectrum of subjects. The paper is an excellent resource on the decade's social climate.

27. Carter's career, and especially his impact as a professional reformer, is covered in Machoian, *William Harding Carter.* Carter used the term "bitter dose" in his memory of events associated with Smith in Memoirs, 44.

28. Carter diary, November 10, 1870.

29. "The Colored Cadet," *New York Times,* January 5, 1871.

30. Carter diary, November 10, 1870.
31. Carter, Memoirs, 5–6. Carter claimed in his memoirs that Ike Carter, his childhood friend, the son of an enslaved couple, had enlisted in one of the Black infantry regiments, hoping to serve with Carter on the frontier, not understanding that the army was strictly segregated and the improbability that their paths would ever intersect. This story is not confirmed, but there also is no reason to doubt Carter's veracity, even if his memories by the time he wrote were biased by nostalgic feelings for his childhood. Carter's mother thought enough of his friendship with Ike to tell him in a letter that she had visited Nashville and seen Ike and that he and his family were well: Carter to Mrs. Samuel J. Carter, April 16, 1871.
32. "West Point May Get a Colored Cadet," *New York Times*, January 24, 1910.
33. For an example entry where Carter bemoans Smith as a liar, see Carter diary, November 10, 1870; quotation from Carter to Mrs. Samuel J. Carter, February 12, 1871.
34. Carter diary, December 26, 1870; and Carter to Mrs. Samuel J. Carter, January 1, 1871. Carter's roommate and Smith were both placed under arrest for several days after the altercation.
35. Pappas, *To the Point*, 379–80. My retelling of the 1871 "vigilante" scandal is based on Pappas's work and Dillard's "Military Academy" as well as editorials and articles appearing in *Army and Navy Journal*. Interestingly, President Grant's son was involved in the episode as one of the informal ringleaders, and some suspect that external pressure on Colonel Pitcher to more completely investigate the events was motivated by those who wished to publicly embarrass the president. There is no mention of this event or of racial integration in West Point's own history of its first one hundred years, *The Centennial of the United States Military Academy at West Point, New York: 1802–1902* (Washington, DC: Government Printing Office, 1904).
36. On the history of West Point's honor system and its eventual codification, see Lewis Sorley, *Honor Bright: History and Origins of the West Point Honor Code and System* (New York: McGraw-Hill, 2009). Sorley makes it clear that very little of real definition is known and much is assumed about the early evolution and implementation of the code.
37. Carter to Mrs. Samuel J. Carter, January 9, 1871.
38. Carter to Mrs. Samuel J. Carter, January 9, 1871. Peter S. Michie (longtime USMA professor and member of the Academic Board) relates the "vigilante" event from the viewpoint of Emory Upton, then commandant of cadets, in *The Life and Letters of Emory Upton* (New York: D. Appleton and Co., 1885), 257–72. This work, largely hagiography, contains no references to the cadet career of James Webster Smith.
39. Carter diary, February 3, 1871. On the "lying scandal" and its relationship to the cadet corps' view of Smith's presence at West Point, see Dillard, "Military Academy," 199–200.
40. Carter to Mrs. Samuel J. Carter, January 9 and February 12, 1871. The previous fall, a *New York Times* correspondent reported that cadets generally viewed Smith's retention as an inconsistent travesty of the USMA code of honor; see "West Point:

Excitement among the Cadets—Indignation at the Recent Decision of the War Department," *New York Times,* November 17, 1870.

41. The court was seated on January 5, 1871, and reached a guilty verdict after hearing several days of testimony. See Dillard, "Military Academy," 197–98; and also the *New York Times* coverage, especially that of January 5, 11, and 13, 1871. The *Army and Navy Journal* (8.22) references the January 1871 court as yet another opportunity for the "political" newspapers' "great noise" over Smith's claimed "oppression."

42. Carter to Mrs. Samuel J. Carter, February 12, 1871.

43. Carter to Mrs. Samuel J. Carter, February 12, 1871.

44. Carter to Mrs. Samuel J. Carter, March 19 and April 2, 1871. Carter relates he has "long prayed" for Cadet Smith's dismissal in the April 2 letter.

45. Carter to Mrs. Samuel J. Carter, April 2, 1871.

46. On charges brought against Smith in March 1871, see Dillard, "Military Academy," 198. Dillard theorizes that the commandant may have been hesitant to pursue court-martial proceedings against Smith by this point due to the Grant administration's demonstrated reluctance to support such action.

47. The text of Secretary Belknap's ruling on Smith's dismissal was published in the *New York Times,* June 15, 1871.

48. Carter diary, June 16, 1871.

49. On reference to Napier's appearance, see Carter diary, June 1 and 8, 1871.

50. "Statement Showing the Number of Colored Persons Appointed Candidates for Admission to the U.S. Military Academy," October 21, 1886, RG 404, USMA Library Special Collections and Archives.

51. Henry O. Flipper, *The Colored Cadet at West Point: Autobiography of Lieut. Henry Ossian Flipper, U.S.A., First Graduate of Color from the U.S. Military Academy* (New York: Homer Lee and Co.,1878; reprint, Lincoln: University of Nebraska Press, 1998), 37 and n37. Flipper's book is an excellent resource on this subject, and he reproduces many of the published letters and news articles pertinent to the Smith case in a last chapter as a sort of addendum. Rory McGovern, Makonen Campbell, and Lisa Koebrich discuss Smith's reaction to white provocations in depth in "'I Hope to Have Justice Done Me or I Can't Get Along Here': James Webster Smith and West Point," *Journal of Military History* 87.4 (October 2023): 964–1003. This article represents the most recent focused scholarship on Smith's time at West Point and is thoroughly documented.

52. On Smith's failure and dismissal, see McGovern, Campbell, and Koebrich, "I Hope to Have Justice Done Me," 1000.

53. Editorial article no. 5, *Hartford Daily Courant,* June 9, 1874; "West Point," *New National Era,* July 30, 1874; editorial comment, *Army and Navy Journal* 11.48 (July 4, 1874).

54. On Smith's last years at West Point and his activities after leaving the academy, see Vaughn, "First Negro Cadet," 101–2. On Smith's public repudiation of his dismissal,

see the *Hartford Daily Courant,* August 13, 1874. Many of Smith's letters to various newspapers are reproduced in Flipper, *Colored Cadet,* 288–308.

55. The story goes that as a cadet, Upton finally fought Wade Hampton Gibbes, a Southern cadet who had insulted him in some particularly offensive manner due to Upton's abolitionist principles. For more on this episode and an insightful biographical study of Upton's career, see David John Fitzpatrick, "Emory Upton: The Misunderstood Reformer" (PhD diss., University of Michigan, 1996). Upton's fight with Gibbes is found on pp. 46–47.

56. Fleming, *Military Academy,* 219. This vignette appeared in "The Colored Cadet," *New National Era,* July 21, 1870. It also ran in New York's *Daily Graphic* and was republished from that source in Flipper, *Colored Cadet,* 319.

57. See McGovern, Campbell, and Koebrich's analysis on this point in "I Hope to Have Justice Done Me," 993–95.

58. Fitzpatrick, "Upton: Misunderstood Reformer," 192.

59. Ron Chernow, *Grant* (New York: Penguin Press, 2017), 769. Chernow contends that Grant's support for Smith was lukewarm or even lost due to his son Fred's assertions that Smith was not right for the academy.

60. "Conduct Becoming Gentlemen," *New National Era,* August 27, 1874. This same issue called on Republican congressmen to deny fiscal support for the academy in retribution for its failure to protect Black cadets.

61. "Bravery and Honor at West Point," *New National Era,* August 13, 1874. Contemporary observers believed President Grant was personally interested in Smith's success: see "The Campaign Opened," *New York Times,* September 12, 1872; although, as noted, Grant's support may have been less than energetic early in Smith's career due to his own son's involvement. The *New York Times* asserted on several occasions that staff and faculty attitudes toward racial integration at West Point were overwhelmingly the same as those of cadets—though not evinced aloud. See, for example, "West Point: Excitement among the Cadets," November 17, 1870; "The Colored Cadet," January 5, 1871; and "West Point: The Case of Cadet Smith," January 6, 1871. Upton's and Pitcher's official support was a matter of public record, and from extant accounts, it seems likely that both men sought objective adherence to the rule of order and discipline regarding Smith's presence, but it is also likely that no coherent and persistent effort was made to effectively modify the corps of cadets' negative reaction. On this topic, see also Rory McGovern's essay in this volume, "You Need Not Think You Are on an Equality with Your Classmates."

62. For representative episodes illustrating Carter's concern for minority outrage, see Machoian, *William Harding Carter,* 46–47; and Carter, "A Plea for the Filipinos," *North American Review* 184 (February 15, 1907): 383–88.

63. George L. Andrews, "West Point and the Colored Cadets," *International Review* 9.5 (November 1880): 477–98; and Peter S. Michie, "Caste at West Point," *North American Review* 103.283 (June 1880): 604–13. Both articles rationalize the general

failure of West Point's "colored" cadets (Smith and Whittaker) in terms of the period's larger social climate and question the logic of having attempted racial integration in the first place. The three early graduates are Henry Ossian Flipper (Class of 1877); John Hanks Alexander (Class of 1887); and Charles Denton Young (Class of 1889). The latter officer later served as military superintendent of Sequoia and General Grant National Parks and as attaché to both Haiti and Liberia; he also accompanied Pershing's Punitive Expedition into Mexico (1916–17). In addition to Flipper's *Colored Cadet,* Wesley Brown follows the careers of these and later African American graduates in "Eleven Men of West Point."

"West Point, New York.—Court-Martial of James W. Smith, the colored cadet—Smith reading his defense, January 12th." This newspaper sketch from *Frank Leslie's Illustrated* shows James Webster Smith defending himself at his second court-martial in 1871. Admitted in 1870 and dismissed in 1874, Smith was the first Black cadet admitted to West Point. His persistence altered the environment for the Black cadets who followed him to the academy. (Courtesy New York Public Library)

Born enslaved in Georgia in 1856, Henry Ossian Flipper became the first Black cadet to graduate from West Point in 1877. (Courtesy United States Military Academy Library Special Collections and Archives)

This stereo card captures Henry O. Flipper in the ranks during a summer encampment, likely in either 1874 or 1876. Flipper stands in the rear rank, nearest the camera. (Courtesy USMA Library Special Collections and Archives)

Like the previous image, this photograph captures Henry O. Flipper at the end of the second rank (*center left, one row back*) in this formation outside the main cadet barracks. These two images depict how isolating it was to be a Black cadet at West Point during the Reconstruction period. (Courtesy USMA Library Special Collections and Archives)

Cadets pose outside a building near the barracks area circa 1866–70. Private spaces without persistent presence of officers and others in authority could pose unique challenges to Black cadets at West Point during Reconstruction. (Courtesy USMA Library Special Collections and Archives)

This photograph shows a typical cadet barracks room in 1870. Rooms accommodated two to three cadets at a time. These were segregated spaces— Black cadets roomed together or alone, depending on how many were at West Point at a given time. If a Black cadet lived alone, these rooms were places of isolation and vulnerability. James W. Smith and Johnson C. Whittaker were both assaulted at least once while in their rooms. (Courtesy USMA Library Special Collections and Archives)

This 1870 photograph shows the cadet library and cadet chapel on the south end of West Point's plain, adjacent to the barracks and academic building. The library and the chapel were each sanctuaries in their own right for Black and white cadets alike. (Courtesy USMA Library Special Collections and Archives)

William Harding Carter graduated with the Class of 1873. His diary and letters reveal much about how White cadets responded to the integration of West Point. (Courtesy USMA Library Special Collections and Archives)

Civil War hero Emory Upton served as commandant of cadets at West Point from 1870–75. A committed abolitionist before the war, Upton took some steps to ensure fair treatment for Black cadets. But his faith in the institution blinded him to the most serious challenges they faced. (Courtesy USMA Library Special Collections and Archives)

Emory Upton (*left*) relaxes in his tent with the post surgeon during an annual summer encampment in the early 1870s. As commandant Upton oversaw the military training and discipline of the corps of cadets. (Courtesy USMA Library Special Collections and Archives)

"Superintendent's Quarters—Colonel and Mrs. Pitcher sitting on front veranda." From 1866 to 1871, Thomas Pitcher served as superintendent at West Point. He was in that position when James W. Smith and Michael Howard arrived in May 1870. Like Emory Upton, he intended to treat Black cadets fairly, but his faith in the institution left him unable or unwilling to perceive and address the most significant issues they faced. (Courtesy USMA Library Special Collections and Archives)

The barracks building circa 1874. The small figure standing to the left of the sally port is longtime West Point bugler Louis Bentz preparing to sound assembly. Bentz is the only individual mentioned by name in accounts from Reconstruction-era Black cadets for having actively and consistently refused to partake in their "silencing." Smith identified him by occupation in letters home, and Flipper named Bentz in his memoir. Both referred to regular conversations with the old bugler as one of the only respites from their isolation. (Courtesy USMA Library Special Collections and Archives)

Cadets and what appears to be one young officer relax on one of the "company streets" during the annual summer encampment at West Point. Observe how close the tents are to one another in the background. It would have been impossible for other cadets to not hear the threats made against James W. Smith during his first night in camp in June 1870, though no cadet would testify to having heard them. (Courtesy USMA Library Special Collections and Archives)

Charles E. S. Wood (*in white*) and other cadet officers in the rising first class (senior class) of 1874 pose for a picture in front of his tent during the annual summer encampment in 1873. James W. Smith entered the academy with the Class of 1874. Records indicate that Wood actively made trouble for Smith in April 1871. (Courtesy USMA Library Special Collections and Archives)

Johnson C. Whittaker (*inset, top left*) was the last Black cadet admitted during Reconstruction, having reported to West Point in August 1876. His assault (1880) and subsequent trials (1880–82), the first of which is depicted in this courtroom sketch, caused a public sensation and inspired some public sympathy for the plight of Black cadets. (Courtesy West Point Museum)

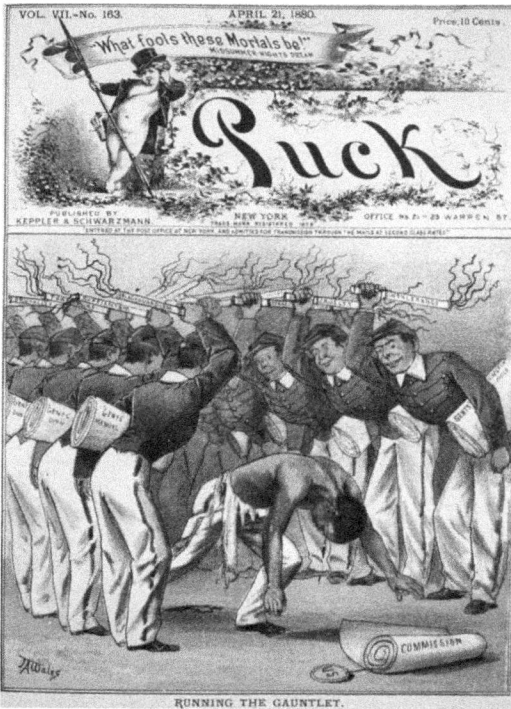

VOL. VII.–No. 163.   APRIL 21, 1880.   Price 10 Cents.

"What fools these Mortals be!"
MIDSUMMER NIGHTS DREAM

Puck

PUBLISHED BY
KEPPLER & SCHWARZMANN

NEW YORK
OFFICE No. 21-23 WARREN ST.

RUNNING THE GAUNTLET.

This cover for the April 21, 1880, issue of *Puck* magazine captures public sentiment turning against West Point for the injustice done to Johnson C. Whittaker, then accused of orchestrating his own assault to escape end-of-year examinations. The imagery of cadets using cat-o'-nine-tails (labeled "Brutality," "Torture," "Ruffianism," "Blackguardism," "Intolerance," "Cruelty," and "Persecution") to whip a Black man pursuing an officer's commission is quite visceral. (Courtesy Library of Congress)

"YOU NEED NOT THINK YOU ARE ON AN EQUALITY WITH YOUR CLASSMATES"

Resistance to Integration at West Point

Rory McGovern

June 29, 1870, was a clear day at West Point. It was the first full day that the newly admitted Class of 1874 was present for the annual summer encampment. Squads of new cadets filled the picturesque plain as the sounds of upperclassmen impatiently barking orders and explaining basic drill and ceremony to their new charges announced the dawn of a new season at West Point. Andrew Howland Russell, then entering his last year at the United States Military Academy and assigned to serve as the cadet quartermaster responsible for supplying the corps of cadets, surveyed the scene from his tent. His attention quickly fixed on one squad nearby, where something seemed amiss. James Webster Smith—the first Black cadet admitted to West Point—stood awkwardly out of formation on the left flank of his squad. Quincy O. Gillmore was berating Smith and ordering him to stand apart from the squad. Then entering his second year at West Point, Cadet Gillmore was the son of Maj. Gen. Quincy A. Gillmore, who had directed the famed assault of the segregated 54th Massachusetts Volunteer Infantry against Battery Wagner outside of Charleston, South Carolina. "You need not think you are on an equality with your classmates," Gillmore proclaimed loudly enough for Russell to hear it as clearly as if he were standing in formation with the squad, "you are not."[1]

By that point, Smith needed no reminders that he had entered a hostile environment. He had already endured a bucket of urine and feces being dumped on him as he slept while awaiting entrance examinations and had spent his first night in camp frozen in terror in his tent as two white cadets stood outside and threatened to blow it up with a barrel of gunpowder. But the episode at drill taught Smith that he could expect neither help nor refuge from fellow cadets and his chain of command. "If you learn anything you have got to learn it yourself," Gillmore explained with a sense of finality, "I will not teach you anything."[2]

Gillmore's behavior that day was patently insubordinate. He was responsible for training a squad of new cadets but refused to accommodate one in particular, whose congressional nomination and successful entrance and medical examinations were no less valid than any of the others in his squad. By principle and by policy, Smith had every right to be at West Point. But Gillmore believed that he could and should deny Smith that right and could act in ways meant to block Smith's path toward a commission. Through such behavior, Gillmore actively resisted the policy of integration at West Point. And though the form of his resistance was extreme enough to be noticed and earn a reprimand, the act of resisting integration placed Gillmore firmly in the mainstream of cadet and faculty responses to integration at West Point.[3]

Cadets and faculty alike resisted integration. Cadet resistance was usually active. White cadets never accepted Black cadets as true members of the corps and demonstrated that rejection in ways ranging from ostracism to violence. Faculty resistance was more often passive and was occasionally the unintentional byproduct of blind faith in the institution they served combined with the subconscious bigotry that was so common in their time and class. This created an environment in which West Point was blind to its own structural flaws and was led by officers and professors unwilling or unable to take meaningful steps to ensure that Black cadets had as viable a path to graduation as their white counterparts. In its relative passivity, faculty resistance was more pernicious, doing much to reinforce and sustain cadet resistance. Cadets perceived all resistance from the faculty, regardless of its type, as a tacit endorsement of their own opposition. This created an environment in which white cadets faced few if any consequences for tormenting Black cadets, and Black cadets had no viable recourse to address wrongs or otherwise ensure fair and just treatment.

By focusing on resistance from cadets, faculty, and officers alike, this essay breaks new ground. With few exceptions, previous scholarship has not studied in any systematic way cadet and faculty responses to integration at West Point during Reconstruction. Most of the literature rightly focuses on the lived experiences of Black cadets, generally sustaining the long-held interpretation that white cadets responded poorly and were sources of considerable friction for their Black classmates while West Point's faculty set aside their individual prejudices and with few exceptions treated Black cadets with fairness and professional propriety.[4] Responses from white cadets and faculty appear not as subjects to examine critically in their own light but as plot points—events and remarks that serve as vehicles to propel the narrative by adding tangible details to Black experiences at West Point.[5] But to better understand integration at West Point in the late nineteenth century, we must also account for how white cadets and faculty responded to the introduction of Black cadets. Those responses intersected with and affected not only Black cadets' experiences at West Point, but also the process of integration itself.

Six Black cadets may have been admitted to West Point during Reconstruction, but the United States Corps of Cadets never accepted them. White cadets tried to compel their first Black classmates to depart. When James Webster Smith's persistence proved the limits of their coercive power, white cadets adopted a somewhat more passive form of resistance meant to isolate Black cadets and maximize their chances of failing out of the academy. Violence underpinned both forms of resistance. At first, violence was extreme, calibrated to terrorize and at times to cause serious harm. As resistance became generally more passive, white cadets adopted new forms of violence meant both to assert their claims of social superiority and to enforce the isolation of Black cadets in ways meant to cause them to fail.

Arriving in May 1870 to await entrance examinations in early June, James Webster Smith and Michael Howard encountered a changing corps of cadets. Still under the watchful eyes of the commanders of various Reconstruction districts, congressional delegations from the last of the formerly seceded southern states had been reinstated by the spring of 1870. Although they were visible manifestations of a new South, Smith and Howard—hailing from South Carolina and Mississippi, respectively—were part of a growing number of West Point cadets coming from states in which the institution of slavery had been legal within their lifetimes. For obvious reasons, there had been a sharp decline in admissions of cadets from slave states during and immediately after the Civil War, as most had been in rebellion. After the war, young men from those states could not secure West Point nominations until properly represented in Congress. Upon admission in 1866, 10 percent of the class of 1870 came from states in which racial slavery was a legal and accepted practice within their own lifetimes. By 1870, however, most states had regained congressional representation. Arriving in the summer of 1870, nearly 25 percent of the Class of 1874 came from states in which slavery had been legal in their lifetimes. By 1875 the proportion of the incoming Class of 1879 from former slaveholding states had risen slightly more, to 27 percent.[6]

The majority of cadets came from free states or states in which slavery had been legal but not central to social and economic life, but this did not make the corps of cadets receptive to integration. Racism was not unique to the South. Historian Eric Foner has argued persuasively that "racism was pervasive in mid-nineteenth-century America and at both the regional and national levels constituted a powerful barrier to change."[7] Moreover, West Point was at its nadir of discipline after the Civil War. The corps of cadets had become infamously prone to vigilantism, and some went to extreme lengths to enforce conformity to unwritten social codes. At the same time, officers in key positions either failed to resolve or even actively encouraged and abetted disciplinary problems in the corps.[8]

Given broader social trends that defined the Reconstruction era, as well as the academy's lax discipline and problematic leadership, it is little surprise that Smith found an unwelcoming environment that summer at West Point. Cadets possessing a variety of views on race and society created a downward spiral defined by consensus around and conformity to the most regressive views on race and military service. Even many of those cadets with family ties to causes of emancipation and civil rights became antagonists to Black cadets. Cadet Frederick Dent Grant—son of President Ulysses S. Grant, whose policies brought Black cadets to West Point—argued that it was not yet the proper time to admit Black men and actively participated in isolating James Webster Smith from the corps of cadets.[9] Similarly, one of Smith's earliest tormentors was Cadet Quincy O. Gillmore, whose father, as mentioned earlier, had commanded the famed 54th Massachusetts and a number of other United States Colored Troops regiments that provided valuable service in the Department of the South during the Civil War. Despite varied backgrounds and the range of views represented within it, a dominant undercurrent of racialized resentment ran through the corps of cadets when two young Black men arrived among them that summer.[10]

Such views festered unmolested as leaders did nothing to prepare the corps of cadets for the arrival of the academy's first Black prospective cadets—those who had received congressional nominations but had not yet taken and the entrance examinations that served as the final gate to admission. This was partially because West Point did not have long to prepare. Cadets usually received their nominations a year or more before reporting to the academy. But the first Black prospective cadet, Michael Howard from Mississippi, disembarked at West Point only one month after receiving his nomination. Mississippi was slow to accept and adhere to conditions for readmission to the Union after the Civil War. Congress did not recognize and seat a delegation from Mississippi until February 1870. Newly seated congressman Legrand Perce nominated

Howard on April 20, 1870. Howard accepted the nomination on April 30 and arrived at West Point on May 25.[11] James W. Smith followed shortly thereafter on May 31, having received his nomination only eight days earlier from Solomon L. Hoge, who had to race to fill his district's unexpected academy vacancy in the spring of 1870.[12]

But even had they more time to prepare, academy leaders were not inclined to take any action to condition the corps of cadets to receive and accept young men of color. Col. Henry M. Black, who served as the commandant of cadets until July 1870 and presided over the corps of cadets when Smith and Howard arrived, later testified to a court of inquiry that he took no "unusual precautions" on behalf of Smith and Howard. He professed to be perfectly satisfied that he and other officers had already taken "every precaution in our power to prevent the ill-treatment of candidates for admission" regardless of race. There is an obvious contradiction between taking no "unusual precautions" and having already taken "every precaution in our power." Had he actually done the latter, it would constitute an unusual precaution, as he did not have to take precautions on behalf of any other prospective cadets. The most charitable explanation is that Colonel Black was unrealistically optimistic; but the clear contradiction suggests that he was simply not inclined to act.[13]

Unrestrained, the corps of cadets responded to Smith and Howard appearing among them with shock and disbelief. A *New York Sun* reporter covering Michael Howard's arrival on May 25 reported that the appearance of the first Black prospective cadet left the academy both "breathless with excitement" and "almost speechless." He noted that white cadets "seemed paralyzed." Surveying their responses to Howard, the *Sun* reporter heard one cadet characterize the situation as "dreadful"; another suggested that they should throw Howard into the Hudson River; others threatened to resign rather than attend an integrated West Point; and still others "talk[ed] of killing the black boy outright."[14]

Initially, white cadets were too shocked to muster any cohesive collective response other than to hurl epithets on those whom they perceived to be new interlopers in their isolated world. In an otherwise entirely optimistic letter, James Webster Smith reported of his first two days at West Point: "The cadets call us [Smith and Howard] niggers, of course. . . . The first day that we were here we could hear nothing else but that word ringing out all sides, from every window, and nook, and niche, continually."[15] Because the white cadets apparently believed that Smith and Howard would not actually be admitted to the academy, this initial response was quite restrained compared to what would soon come. Many cadets assumed that officials would find a reason to reject Smith and Howard during their final medical and academic examinations before admission. The *New York Sun* revealed as much in its reporting, which predicted that Howard would pass the medical exam, but "then he will fail the mental examination, and go back to Mississippi" because "the examining officers have power to reject any applicant." Both the inspector of the academy and the commandant of cadets "are opposed to the African," according to the *Sun*, "and while they are at the head of the National Academy, the black boy will remain on the plantation."[16]

But as examinations approached, cadets and prospective cadets alike attempted to push Smith and Howard to quit before the examinations rather than take the risk that one or both would be admitted. Those who arrived with Howard and Smith were the first to act. On June 7, 1870, prospective cadet Robert McChord of Kentucky pushed roughly through a doorway that he thought was obstructed by Smith and Howard, shouldering Howard and demanding that he and Smith move. On their refusal, McChord slapped Howard across his face, yelling at him to get out of the doorway. Outraged, Smith challenged McChord. Enraged by Smith's boldness, McChord reached for his pocketknife and growled, "I ought to cut you open." Though he did not pull the knife out, the threat was clear as several prospective cadets in the room crowded in, giving voice to simmering animosity for Smith and Howard with shouts of "cut them open," and "kill the d——d niggers."[17] Cooler heads eventually prevailed, averting additional violence that day. But the rapidity and extremity of the white responses to an altercation between Smith and McChord were revealing. Mob action cannot happen in the absence of an apparent consensus. Through their vicious threats, white prospective cadets signaled that they would not accept Smith and Howard as part of the larger group and that they believed violence was an acceptable way to preserve an all-white corps of cadets.

They reinforced and amplified that message in any way possible as the medical and academic exams loomed closer. Some snuck into Smith and Howard's room a few nights later and dumped the contents of a slop pail—a bucket that in the absence of indoor plumbing collected all kinds of waste products and byproducts throughout the day and night—onto Smith and Howard as they slept.[18] In a letter to his benefactor, Smith observed that "the cadets (especially the new ones) are down on us," and that when he and Howard were "put into the squad to drill, one of the white boys (Crane from Ohio) refused to drill with us."[19] Beyond refusing to drill, white cadets and prospective cadets went so far as to deny their Black comrades food. In a bold power play at the mess table, Crane—the same prospective cadet who attempted to refuse to drill with Smith and Howard—and several others insisted upon serving themselves before Smith and Howard could eat, and then passed all the food to the other end of the table, well out of their reach. The cadet officer in charge of that mess table refused to intervene. Smith wanted to report the matter, but Howard prevailed upon him that "if we complained about it, they would think we were greedy." According to Smith, the situation at the mess table got so bad that "what I get to eat I must snatch for like a dog." When asked about the truth of Smith's statement, Crane laughed it all off as something "of the nature of a joke."[20]

Efforts to drive away James W. Smith only escalated after he was formally admitted to the academy. While the Academic Board denied Michael Howard admission for failing the academic examination, Smith sailed through. Conspicuously alone, he became a target for unique torment. He spent a sleepless first night in his tent frozen in terror as two cadets stood outside loudly discussing plans to place gunpowder under the wooden platform it was pitched upon, debating when to light a fuse and from where to watch the explosion. They went so far as to raise the floor on one side of the tent several inches, pretended to shove gunpowder underneath it, and yelled to each other to ignite it. "I did not sleep two hours all night," Smith reported.[21] This went far beyond the hazing that befell most new cadets. However badly conceived, hazing served as a perverse welcoming ritual through which cadets subjected new cadets to annoyance, inconvenience, and humiliation to initiate them to West Point.[22] Instead, Smith faced terror and deliberate isolation imposed by those who intended to compel either his resignation or his dismissal. Things did not improve for Smith when training began. The incident described at the beginning of this chapter in which Quincy O. Gillmore refused to train Smith as part of his squad happened on Smith's first full day in the summer encampment as an admitted cadet.[23]

By separating Smith from the squad and refusing to train him, Gillmore's actions were calibrated more to force Smith out of the academy than to enforce segregation or otherwise lay claim to social superiority. Without training, Smith would make mistakes at drill. Mistakes at drill could lead to demerits. And demerits could lead to expulsion—more than one hundred in any given six-month span would prompt the Academic Board to declare a cadet deficient in discipline and to dismiss that cadet from the academy. Ignorance of regulations was therefore just as much a threat to cadets' careers as poor academic performance. By isolating and refusing to train Smith, Gillmore not only protested Smith's admission to the academy but also demonstrated to other cadets a way to force his expulsion. White cadets took notice. Isolation became a central pillar of their response to integration at West Point, starting with James W. Smith. Writing home to his brother that July, Cadet James Fornance said in reference to Smith that "no-one speaks to him nor has anything more to do with him than possible."[24] In the same month, Cadet Charles Wooden reported that Smith "has been left almost entirely to himself, rarely being spoken to by cadets except on duty."[25] The corps of cadets had collectively decided to "silence" Smith.

At the time, it was common practice for the corps of cadets to silence or "cut" cadets who were seen as dishonorable or as bringing dishonor on the rest of the corps. Silencing was social ostracism—cadets would refuse to speak to a silenced cadet other than to communicate orders as required while carrying out official duties. Most commonly, this fate befell cadets suspected of lying or, on the other side of the spectrum, of cooperating too much and speaking too freely when officers investigated offenses of other cadets.[26] Silencing was a form of ostracism and isolation that often resulted in resignation or dismissal due to academic or disciplinary deficiencies. Cadets invariably relied on the assistance of other cadets to navigate West Point, whether for assistance in studying and understanding a subject for which they had little natural aptitude or affinity— borrowing uniform items needed to get through drill or guard duty, or gaining help in cleaning and maintaining barracks rooms, uniforms, and equipment. Silenced cadets received no assistance. If the social strain did not cause silenced cadets to depart the academy voluntarily, subsequent academic or disciplinary deficiencies often forced them out.[27] But waiting for silencing and isolation to drive Smith away would take more time than many cadets were willing to give.

Instead, some reacted violently. In Smith's first year that violence was extreme, calibrated either to terrorize Smith enough that he would resign or to inflict serious physical harm. White cadets had threatened violence against Smith and Howard in their earliest days at the academy. In his altercation with Smith, Robert McChord had threatened to stab Smith, prompting his entire squad of prospective cadets to crowd around and urge him to stab both Smith and Howard. And then two cadets threated to blow up Smith's tent during his first night in camp after passing the entrance examinations. Violence toward Smith was a very real and persistent threat, compounding the anxiety of his every hour.[28]

Eventually, white cadets acted on their threats. When attempting to collect water from a well on August 13, 1870, Smith endured a brutal fight in which both he and his assailant exchanged blows with heavy wooden ladles meant for scooping water into and out of a bucket.[29] Later, when Smith helped himself to soup at the mess table before one of his white messmates had taken a share, the white cadet was so incensed that he followed Smith into his barracks room and viciously attacked him while other cadets blocked the hallway, held the door shut to prevent Smith's escape, cheered, and shouted encouragement for their friend to "kill the d——d nigger." With his assailant going for the throat, Smith understandably sensed mortal danger and resorted to grabbing a bayonet from the equipment in his room to force his attacker to stop and flee.[30] Violence against Smith was common enough that Cadet William Harding Carter's diary entry for November 10, 1870, opens with a statement all the more disturbing because of its chilling casualness and banality: "My roommate has just skinned the 'nigger.'"[31] In the vernacular common to cadets at the time, "skinned" could have referred to hazing in the form of making lower-ranking cadets do inconvenient, painful, or absurd things. But in transcripts of Smith's three separate court proceedings in 1870–71, dozens of cadet witnesses uniformly insisted that they refused to haze Smith in the same manner that they hazed white cadets either because they feared extra scrutiny and punishments or because doing so would put Smith on a certain level of equality with his peers.[32] In this context, then, Carter's statement most likely referred to a physical assault, consistent with another definition of "skinned" common in the late nineteenth century. Violence had become a routine matter in white cadets' interactions with Smith, and it had escalated far beyond petty violence intended to assert superiority or enforce social norms.

Simultaneously, cadets actively sought out other methods to ensure Smith's dismissal if he did not resign. One such method amounted to hunting for disciplinary demerits. As noted earlier, cadets who accumulated more than a hundred demerits in any six-month period faced dismissal for deficiencies in discipline. Demerits could be awarded for any number of reasons.[33] White cadets strove to ensure Smith would receive more than his share. They scuffed his shoes, creased his uniform, stepped on his feet to trip him during drill, and reported him for smiling in the ranks or for other, more vague charges of inattention, talking, or disrespect in the ranks.[34] While Smith did not lead his class in demerits, those he did receive came almost entirely from reports and actions of fellow cadets. White cadets at the time were hesitant to report offenses against each other, unwilling to be seen as bringing undue trouble upon fellow cadets or as cooperating too readily with officers' investigations against fellow cadets. To do so risked being silenced by the rest of the corps while at the academy or even continued ostracism well after graduation.[35] White cadets freely and frequently reported Smith's real and imagined infractions and otherwise tried to create infractions for officers to notice. They were attempting to weaponize the demerit system against Smith, in the process usurping one of the methods used to instill military discipline.

White cadets also tried to use the military justice system to force Smith out of the academy. Lying was a gravely serious offense at West Point, one that could result in dismissal. Paradoxically, however, many white cadets proved willing to perjure themselves to make it appear that Smith had been caught in a lie, hoping that he would be charged, found guilty, and summarily dismissed.

This usually manifested as a conspiracy of silence whenever Smith raised allegations of mistreatment or wrongdoing. Smith's accusations and rebuttals
invariably prompted white cadets to deny and dissemble. And there was a
certain deliberateness to their denials. Throughout surviving records, white
cadets almost always couched their rebuttals of Smith's claims with careful
qualifiers such as "I cannot say positively," "I do not remember exactly," or "to
the best of my recollection"—all seemingly meant to safeguard against future
charges of perjury.[36] When Smith alleged he had been mistreated, consistent
denials from those he said mistreated him and those assumed to have witnessed the mistreatment led investigating officers to charge Smith with making false statements and with conduct unbecoming an officer in the United
States Army.[37] Cadets were only too willing to refute Smith under oath in the
hopes that doing so would prove Smith a liar and cause his expulsion. In one
of the more blatant examples of this phenomenon, a cadet accused by Smith of
mistreatment denied everything but admitted under oath that he had spoken
to other witnesses before the trial began and freely stated that he hoped the
trial would prove Smith to be a liar and lead to his Smith's expulsion.[38]

It nearly worked. Smith spent much of his first year under investigation and
arrest, and each episode exacerbated the animosity and contempt other cadets
held for him. He faced one court of inquiry and two courts-martial in his first
eight months at the academy. In each, the courts ruled against him, but War
Department or Grant administration officials modified the recommendations
and sentences. The court of inquiry in July 1870 recommended that Smith face
a court-martial for dishonorable conduct, a recommendation the secretary of
war rejected, ordering instead a reprimand from the superintendent. Although
the secretary of war had disapproved the court's recommendation, cadets and
faculty alike sided with the court and labeled Smith a liar. This informed their
attempts in two later courts-martial to frame testimony in ways meant to prod
the courts to dismiss Smith.[39] They were further incensed when the first court-
martial found Smith guilty in the spirit of the charges brought against him but
had to impose a light sentence because those who had accused Smith of acting disrespectfully in the ranks had identified the wrong date in the charges,
and the person Smith had been accused of disrespecting had been on guard
duty and not at drill on the date identified in the charges.[40] The second court-
martial in January 1871 found Smith guilty of yet another series of charges and
actually sentenced him to be dismissed from the academy. But after months
of deliberation, the Grant administration determined that the conduct of the
trial was questionable at best and decided that "the ends of public justice will
be better subserved, and the policy of the Government—of which the presence of this Cadet in the Military Academy is a signal illustration—be better

maintained, by a commutation of the sentence." In the end, Smith had to repeat his first year of studies rather than leave the academy.[41]

It would be hard to overstate white cadets' anger over the decision to retain Smith. Alongside anger, however, there was in many white cadets a growing recognition that they could not simply drive Smith or any other Black cadet away. Even before the news that the War Department had modified the sentence from the second court-martial reached West Point, cadets noted that Smith had been "tried so often and the sentences so smothered down in Washington that we have given up all hopes." Changing tack, Carter and other white cadets stopped actively trying to drive Smith away from the academy on the near term and instead doubled down on isolating Smith, to whom they began to refer—coldly and tellingly—as "the anomaly."[42]

James W. Smith's resilience and the Grant administration's willingness to sustain him had therefore fundamentally changed white cadets' response to integration at West Point. The altered response changed the environment Black cadets entered in ways that made graduation possible if still unlikely. By focusing above all on isolation, white cadets adopted a more passive form of resistance that needed months or years to have the intended effect. While this by no means created a favorable environment for Black cadets, it did alter conditions so that those who followed Smith would not experience the extreme violence that he had faced. Nor would white cadets seek to force expulsion through disciplinary demerits or by manipulating the military justice system. Isolation, ostracism, torment, and some lesser degree of violence remained, but in his persistence, Smith had opened the narrowest of trails for others to follow toward graduation and a commission in the US Army.[43]

White cadets continued to resist integration, centering their resistance upon isolation and ostracization. Silencing was the principal technique white cadets used to isolate Black cadets. As noted above, silencing was the practice of refusing to speak to a targeted cadet unless absolutely necessary in the course of official duties. In his last three years at West Point, Smith enjoyed little to no social intercourse. Upon his own arrival at West Point in 1873, Henry O. Flipper recalled receiving a letter from Smith full of advice. While he felt reassured by the simple fact that Smith was still at the academy, what struck him the most was that "it was a sad letter" that related "sad experience" in an exceptionally "melancholy tone." Silencing was torture for a young person already immersed in an austere environment, and Flipper observed how much it had affected Smith, who had by that point endured it for three years. It affected Flipper, too, who recalled in his memoir, "There was no society for me to enjoy—no friends, male or female, for me to visit, or with whom I could have any social intercourse, so absolute was my isolation."[44]

Importantly, however, white cadets placed such emphasis on silencing because its wide-ranging effects posed a grave threat to Black cadets' paths to
graduation—it was more than simply the severe impact of social isolation.
Cadets depended on other cadets to survive at West Point. No such assistance
would be available to Smith when his performance in the classroom began to
spiral, and he eventually departed the academy in June 1874 after failing the
same course many white cadets had struggled through with the assistance of
friends. In silencing their Black classmates, failure was the ultimate goal white
cadets had in mind. Referring to Johnson C. Whittaker—a silenced Black
cadet in his class—Cadet George W. Goethals wrote in a May 1877 letter to a
friend, "Mr. W. is in the 7th Section in Math and we are in hopes that he will be
found" (i.e., failed). The seventh section was the last section, and cadets were
moved between sections according to their performance in the course—the
last section was full of those at risk of failing. Goethals's letter makes it equally
clear that his class passively hoped and waited for Whittaker to fail and that no
one was helping him avoid that outcome.[45] Of the six Black cadets who passed
entrance examinations and gained full admission to West Point throughout
Reconstruction, five were dismissed for academic deficiency after failing one
or more courses.[46] In all of these cases, the outcome was at least partially due
to silencing.

Beyond silencing, white cadets ostracized Black cadets among them to enforce an unwritten social code defined by racial segregation, which thoroughly
contradicted and undermined the policy of integration. However reluctantly
the corps of cadets had come to accept that they could not simply drive away
Black cadets, they never accepted that Black cadets occupied an equal position
within the corps, and they actively worked to build and maintain an informal
caste system at West Point. Some of the more extreme forms of resistance
during James W. Smith's first year at the academy were fueled by perceptions
that Smith represented a real claim to social equality, not only for himself
but also for the broader African American community. When asked during
one of Cadet Smith's several court proceedings whether he had any prejudices
against people of color that would prevent him "from treating them with justice and fairness," Cadet Edward Hardin succinctly stated the representative
view: "I do not know what you mean by treating them with justice. . . . I do not
think they should be made cadets or put on an equality with white people."[47]

Accordingly, wherever they had the power to do so, white cadets denied
Black cadets' participation in social diversions at West Point. For example,
with both James W. Smith and Henry Alonzo Napier at the academy in the
summer of 1871, white cadets grew concerned that the two Black cadets would
attempt to attend the famous summer "hops"—dances hosted by the academy

for cadets and their guests. In separate class meetings, each class agreed that if Smith and Napier appeared at any of the hops, the "hop managers" would promptly announce the end of the event and dismiss the musicians. Smith and Napier kept their distance for a few weeks, but white cadets sensed that would soon change when Smith's brother, sister, and two of his sister's friends arrived to visit Smith and presumably also to attend that weekend's hop. They acted quickly to prevent such mixing. An upperclassman ordered Smith to tie down part of a tent, and once completed, another cadet undid his work. The upperclassmen reported to the officer of the day that Smith had responded disrespectfully and disobeyed orders, prompting that officer to place Smith under arrest. A hasty investigation ruled against Smith and kept him in arrest for five weeks—confining him to his tent other than for drill and meals, thus keeping him from seeing his guests or attending the summer hops and also deterring Napier from attempting to attend on his own.[48]

White cadets sustained a racial caste system at West Point throughout Reconstruction in the usual ways. They first had to unite around, or at least enforce, a common conception of racial superiority. That was no small feat— West Point drew cadets from various backgrounds and from all across the country. But as new cadets became acclimated to life at the academy, they also became acclimated to racial prejudice and bias. Observing his classmates metamorphose pained Henry O. Flipper, who later wrote in his memoir:

> When I was a plebe those of us who lived on the same floor of barracks visited each other, borrowed books, heard each other recite when preparing for examination, and were really on most intimate terms. But alas! In less than a month they learned to call me "nigger," and ceased altogether to visit me. We did the Point together, shared with each other whatever we purchased at the sutler's, and knew not what prejudice was. Alas! We were soon to be informed! In camp, brought into close contact with the old cadets, these once friends discovered that they were prejudiced, and learned to abhor even the presence or sight of a "d—d nigger."[49]

While the archives are generally silent on the process by which white cadets built and sustained a common sense of racial superiority and bias, evidence suggests threats and satire played key roles. The threats were obvious. Cadets knew they could not help Black cadets or break their isolation without risking silencing and isolation themselves. This dynamic becomes evident in records related to Smith's court proceedings. Most of Smith's claims ended up being corroborated by at least one white cadet. But those who did corroborate anything Smith said usually corroborated only one item, and then, as if realizing their mistake, suddenly became unable to remember anything else.[50] Less obvious was the role of satire as a unifying force. White cadets used satire to unite around a sense of racial superiority. Deficiency records note that at artillery drill one day in 1873, Cadet William Davies was reprimanded after he had "disfigured his face" with black powder from the artillery charges.[51] It is difficult to interpret this event as anything other than a blackface incident. More concretely, at some point between 1874 and 1876, Cadet Carver Howland of Providence, Rhode Island, composed a doggerel about James W. Smith and performed it in blackface, strumming a broom as though it were a banjo—a one-man minstrel show. Cadets loved Howland for his many songs and performances. But at least one contemporary observer noted that this particular composition was "his famous and most popular song."[52] Such episodes contributed to the establishment and maintenance of a common bias and sense of racial superiority.

White cadets also asserted that sense of superiority through acts of petty violence. Unlike the violence meant to cause serious physical harm that James W. Smith endured in his first year, the violence that Black cadets endured from 1871 through the end of Reconstruction seemed more calibrated to establish and maintain a social hierarchy. Henry O. Flipper—James W. Smith's roommate during the 1873–74 academic year—recalled being advised by Smith in the summer of 1873 "not to fear any blows or insults," implying both that Flipper should expect violence and that the level of violence was not intolerable.[53] In 1876 Johnson C. Whittaker, the sixth Black cadet admitted to West Point, suffered a blow to the face hard enough to draw blood but not hard enough to cause lasting physical harm by a cadet who took umbrage over Whittaker stepping in front of him in line.[54] And in a letter to Flipper, Whittaker celebrated what he perceived to be improved treatment at West Point because he had suffered only one physical assault in the month of December 1877—an absurd statement if it had come from one of his white counterparts.[55] White cadets clearly threatened and selectively applied violence to assert and enforce a racialized social hierarchy.

Throughout Reconstruction, the corps of cadets' response to integration shifted from active to passive resistance. Once convinced that they could not drive Black cadets away from the academy, white cadets adopted a more passive form of resistance defined by isolation and rigid enforcement of a racial hierarchy to minimize the likelihood that Black cadets would graduate. The white cadets' unwillingness and inability to accept and adhere to the spirit of the policy of integrating West Point sprang from many influences, not least of which were the racial norms and prejudices of their time. But the most proximate influence was the less overt but still visible resistance from West Point's officers and faculty. These leaders were ideally positioned to force cadets to abide by both the letter and the spirit of the policy of integration, but they failed at every turn to do so.

Unfortunately, officers and professors at West Point during Reconstruction also resisted the policy of integration. They were not the paragons of unbiased professional virtue that Henry O. Flipper reported them to be in his memoir. Flipper told the world that "the officers of the institution have never, so far as I can say, shown any prejudice at all" and that "they have treated me with uniform courtesy and impartiality."[56] Readers of his memoir should exercise caution when considering Flipper's opinions of West Point faculty. At the time he wrote, he was still a newly commissioned officer who had every reason to believe a full career was ahead of him. Publishing a book that said anything else about the faculty at West Point would have ended his career since the academy was the heart and soul of the long-serving regular army officer corps. While many of the officers and professors at West Point would have believed, and certainly wanted the public to believe, that Flipper's portrayal accurately described their response to integration, they were in fact almost uniformly prejudiced and partial. And their prejudices and partialities congealed into a pernicious form of resistance to the policy of integration.

    West Point in the 1870s was not a place where new ideas thrived. Reconstruction-era cadets learned from a mostly stable nucleus of long-serving professors in an academy led by officers who in some cases had been taught as young cadets by the very same professors. Three long-serving professors died or departed the academy in 1871, and their average tenure at West Point was thirty-four years. Another departed in 1876 after serving forty-two years at the academy. This constituted a major turnover—prior to 1871, the last departure of one of West Point's Academic Board members had been in 1857.[57] But the replacements were long-serving assistant professors or officers handpicked by departing professors who wanted their former star students to carry on their work. New professors were inclined to carry on in the footsteps of their mentors and to stay on for multidecade tenures at the academy as well. Turnover in this period reinforced rather than departed from past values and practices.[58]

    This created an environment in which those who led West Point were institutionalists who resented the policy of integration as an intrusion on their professional authority and autonomy. In 1880 George Andrews—West Point's long-serving professor of French—published an essay in which he attempted to defend the academy from mounting public criticism in the wake of the infamous assault against Johnson C. Whittaker. In the essay, Andrews related his view of the effort to integrate West Point throughout the 1870s, which he characterized as an unjust drive for complete social equality. He made it clear that he was not impressed with the caliber of Black cadets he had encountered thus far. "It would be well both for the academy and for the colored cadets," he wrote, "that the latter should be of higher character and greater ability than nearly all of his race who have been hitherto sent here." Continuing, he openly questioned the policy that sent Black cadets to West Point. "That the people of the United States," Andrews mused in a flagrant breach of military professionalism, "will sanction the tyrannical course of forcing upon the white cadet an association for which, as their [the Black cadets'] own conduct shows, they are themselves not yet prepared, is not probable."[59] Writing in the same year and for the same reason, Peter S. Michie, professor of natural and experimental philosophy, implied in a different publication that Black cadets did not belong at West Point because those who gained admission prior to 1880 "all displayed a marked deficiency in deductive reasoning, and have taken very low rank in mathematical subjects, but generally possess excellent memories."[60] Even ten years into the effort to integrate West Point, its long-serving faculty were not only unwilling to support the policy but also perfectly willing to question and condemn the policy in public forums.

Open condemnation of government policy was a particularly public form of resistance, but it was not the only way in which those who led West Point resisted the policy of integration. Faculty and tactical officers alike resisted integration in more complex and sometimes subtle ways within more private confines. Their resistance was multifaceted, taking three forms: active, passive, and unintentional. Active resistance on the part of faculty and tactical officers took the form of conscious miscarriages of investigations into mistreatment of Black cadets or abuse of command authority in ways that aided and abetted white cadets' resistance to integration.

Ample evidence proves that James W. Smith suffered from such active resistance from tactical officers. One notable instance was when Smith and Michael Howard filed a complaint after Robert McChord threatened to stab Smith, much to the delight and encouragement of the rest of their squad, for defending Howard after McChord hit him for standing in his way in a doorway. The investigating officer was Capt. A. S. Clarke. Although his investigation found ample corroborating evidence of Smith's and Howard's claims, Clarke ultimately reported that he was "of the opinion that both of these youths [Smith and Howard] have greatly exaggerated an insignificant affair" and conspired with each other to cause trouble for the academy. Tellingly, Clarke confirmed that McChord had assaulted Howard but argued that the greater offense in the whole matter was when Howard attempted to maintain a position in line before a white cadet. Due to a minor discrepancy between Smith's and Howard's statements, Clarke refused to "regard Mr. Howard's testimony as at all worthy of credence." He characterized the assault on Howard as "the inevitable result of his attempting to push himself in the way of his white comrades."[61] Later, that same Captain Clarke was on the case when Smith reported that he and Howard awoke to human waste being thrown on them from a slop pail. While there is a record in the archives that claims Clarke had investigated the matter and "could not fix it upon anyone," no record of that investigation actually exists. As the archives at West Point are replete with Reconstruction-era records of investigations into matters large and small, Clarke most likely reported the results of an investigation that he never actually conducted.[62]

Such miscarriages of justice weren't limited to one tactical officer. During the episode in which white cadets made spurious charges against Smith to cause his arrest and ensure he and his guests did not attend that weekend's hop, Lt. Charles King did not even bother to claim to investigate the matter before arresting Smith. Instead, he acted immediately after hearing from the white cadets.[63] This episode happened after Smith's first year, revealing that such active forms of resistance from at least some faculty and leaders at West Point outlasted active resistance from white cadets, which generally dwindled after Smith's first year.

Others adopted a more passive form of resistance to the policy of integration. Many faculty were able to keep their biases filtered enough to interact with Black cadets courteously and with propriety. But their biases still affected their judgment and treatment of Black cadets. Officers and faculty tended to remain aloof from cadets, concerned that too close an association would cause trouble by allowing that cadet to be resented and labeled a "bootlicker" by the rest of the corps. All the same, cadets could and occasionally did visit officers' and professors' homes to socialize or enjoy a meal, but only if they had a permit or an invitation. No evidence exists of a Black cadet receiving any such invitation. The two who—by remaining at West Point during Reconstruction for more than one year—had the greatest likelihood of receiving such an invitation certainly did not. Smith flatly stated that he had no interactions with officers or faculty outside of official duties.[64] Flipper's memoir includes no stories of visiting an officer's or professor's home because he almost certainly never did. And while Flipper was prone throughout his memoir to praise the faculty for treating him courteously, he pointedly did not include officers or faculty members on the revealingly short list of people with whom "he could and did have a pleasant chat every day."[65] It appears, then, that the officers and faculty at the academy participated in the practice of silencing Black cadets, though it is unclear if they did so with the same goal in mind as the cadets.

More immediately, bias affected faculty members' assessments of the cadets in their charge in ways that constituted passive resistance to the policy
of integration. In their 1880 articles, Professor George Andrews argued that
Black cadets were inherently unequal to white cadets and implied that they
had no place at the academy, while Professor Peter Michie commented on the
"marked deficiency in deductive reasoning" and comparatively poor math-
ematical aptitude he perceived in Black cadets he had encountered. If professors assumed a lack of aptitude and ability in Black cadets, it stands to reason
that they were more likely to believe they observed it during examinations.
Andrews was the professor of French, and Michie was the professor of natural
and experimental philosophy—a subject somewhat similar to what a modern
reader would call physics. Four out of five Black cadets admitted during Reconstruction and dismissed for academic deficiency after at least one semester at West Point faced dismissal for failing examinations in either French or
natural and experimental philosophy.[66]

Beyond active and passive resistance, institutionalism also led some officers
and professors toward unintentional resistance to integration at West Point.
A blind faith in the academy rendered them equally blind to the structural
impediments preventing Black cadets from graduating and to their own role
in building and sustaining those impediments. Emory Upton, who served as
commandant of cadets from 1870–75, is a representative example of this phenomenon. A confirmed abolitionist before the Civil War whose views never
moderated, Upton's appointment as commandant of cadets should have done
much to ease and advance the process of integrating West Point. And Upton
had every intention of being that officer. James W. Smith testified that in his
first days as commandant, Upton had "told me that if I wished to find out anything or make any complaints to come to him as a friend and he would always
see that I was justified in everything that was right." Upton had also addressed
the corps of cadets to remind them that Smith had earned the right to be at
West Point and must be treated fairly by them.[67] But Upton was also a nearly
fanatical institutionalist. He believed that West Point cadets clung firmly to
"the principle that a cadet's word is to be unquestioned" and that "the perfect
trust that exists among comrades, their faith in one another's word, the reli-
ance on one another's charitable assistance in distress, all serve to give this trait
a heathy growth and a real existence."[68] He had such faith in West Point and
the cadets in his charge that he could not fathom that any of them would de-
liberately mistreat a fellow cadet or lie when asked if they had done so.

Accordingly, Upton came to view investigations of maltreatment directed
against Black cadets as simple math problems. Black cadets stated one version of events and multiple white cadets offered contradictory accounts. To

Upton, multiple accounts from white cadets outweighed a single account from a Black cadet. Lacking corroboration, Black cadets were then charged with making false statements in their original accusations, a serious charge that could lead to dismissal from the academy. Upton was unwilling to consider even the possibility that white cadets colluded to keep their stories consistent. This pattern of events happened so often with James W. Smith that Upton felt compelled to preemptively defend himself to the superintendent, opening one letter preferring new charges against Smith with a revealing statement: "The frequency with which I have been compelled to prefer charges against Cadet Smith, J.W. render it necessary that I should state fully for the information of the Superintendent the origin and cause of the charges now preferred." Upton then explained that Smith had accused three white members of his squad of harassing him in formation—but upon separate questioning all three of the accused had stated that it was Smith who had harassed them, leaving him no choice but to charge Smith with conduct unbecoming an officer and a gentlemen and of making false statements.[69] This repeated a pattern he had established in earlier investigations and would carry forward into the future.[70] Upton by no means intended to resist the policy of integration at West Point and on a personal level, may very well have supported it. But his rigid and blind faith as an institutionalist predisposed him to believe that cadets—at least the white ones—were too honorable to lie. This led him to facilitate and accelerate one of the most significant obstacles to the policy of integration: cadets' manipulation of the military justice system.

Smith and subsequently matriculating Black cadets were well aware of this dynamic. Faculty resistance of all kinds—active, passive, and unintentional alike—led Black cadets to lose faith and stop seeking redress when white cadets mistreated them. In his January 1871 court-martial, Smith observed that he never would have defended himself against an accusation of being disruptive in the ranks by noting that a white cadet had stepped on his toes if he had anticipated that the white cadet would deny it. He knew that he would find no cadet to corroborate his story, and so it would have been a hopeless case.[71] Smith also seems to have warned other Black cadets to avoid reporting incidents to their officers if at all possible, in hopes that they would avoid the cycle of recrimination that he endured. In the letter that Flipper received from Smith on his first day at West Point, he was warned to "avoid any forward conduct if [he] wished also to avoid certain circumstances." Flipper made clear throughout his memoir that he held a broad definition of "forward conduct" that included reporting mistreatment. He interpreted Smith's letter as "the confession of some great error made by him at some previous time, and of its sadder consequences." Concluding that seeking help only created more trouble, Smith apparently advised Flipper—and presumably also Napier and Gibbs before him—to make no waves at West Point and to remain singularly focused on his own studies and conduct in order to survive there. The letter had its intended effect. "I don't think any thing has so affected me or so influenced my conduct at West Point," Flipper recalled.[72]

Faculty resistance also influenced white cadets' behavior in response to the effort to integrate West Point. It led white cadets to assume the officers and faculty at West Point tacitly approved their own resistance, making them more confident in the act of resisting. Cadets had no incentive to abandon silencing while their officers and faculty had fallen into a more informal version of the same practice. Moreover, some cadets viewed certain officers and faculty as confederates in the same cause, rooting for and in some cases causing Black cadets to fail out of the academy. Carver Howland's famous doggerel about James W. Smith included the lines:

> Old Davy's done it for you now
> Dis Nigga he has found
> And you won't see any more of this yere childe
> . . .
> G— D— old Lyle.

Those lyrics appear in Eben Swift's unpublished memoir after another song in which it becomes clear that the "Davy" mentioned in any of Howland's songs was Lt. David S. Lyle, who was then assigned to West Point as Michie's assistant professor of natural and experimental philosophy. Howland's use of the word "you" in this verse clearly refers to the broader corps of cadets, while "he has found" refers to Smith failing his natural and experimental philosophy exam—to be "found" was a phrase in common usage at West Point at the time, a shortened form of "found deficient," which meant to have failed an academic examination. Howland's construction of this verse is quite revealing, particularly in the line "Old Davy's done it for you now / Dis Nigga he has found."[73] It reveals that white cadets believed, and celebrated in verse, that Lyle had exercised a controlling hand in Smith's exam result and that he had somehow acted on their behalf. At the very least, it reveals a perception of a united front of white cadets, officers, and faculty to resist integration at West Point.

Throughout Reconstruction, white cadets, officers, and faculty resisted integration at West Point in ways that were distinct but symbiotic. Cadet resistance was active and overt at first. It is unlikely we will ever be able to truly comprehend what James Webster Smith suffered as the first Black cadet at

West Point. Throughout his first year, white cadets found myriad different ways to pursue the goal of driving him out of the academy as quickly as possible, almost as if vying to best each other in a morbid race to the bottom. But Smith endured, and his endurance altered the environment. Realizing after a full year of trying their utmost that they could not drive him away, white cadets then adopted a passive form of resistance calibrated to set conditions for his eventual failure and dismissal. Although Smith ultimately did fail, the altered environment he shaped made it possible if not probable that Black cadets arriving after him would graduate. Faculty and officer resistance ranged the full spectrum—active, passive, and unintentional. It was almost always considerably more subtle than cadet resistance, but it was also more impactful over the long term because it simultaneously complemented, reinforced, and sustained white cadets' resistance. West Point's faculty and officers had more than enough power to force cadets to abide by the policy of integration. But cadets saw through the subtlety. Detecting faculty and officer resistance, cadets felt more secure in their own. Resistance to integration at West Point, then, was as much a failure of leadership as it was a failure of common decency and humanity.

Notes

1. Testimony of Andrew Howland Russell, Untitled Transcript of Court of Inquiry Addressed to Thomas M. Vincent, Assistant Adjutant General (hereafter AAG), July 18, 1870, p. 93, Roll 1, Selected Documents Relating to Blacks Nominated for Appointment to the United States Military Academy during the Nineteenth Century, 1870–1887 (hereafter cited as M1002), United States Military Academy Special Collections and Archives, West Point, NY (hereafter cited as USMA). See also James W. Smith to Friend [David A. Clark], June 29, 1870, included as Exhibit F to Court of Inquiry Transcript addressed to Thomas M. Vincent, ibid.

2. Russell testimony, Untitled Transcript of Court of Inquiry Addressed to Thomas M. Vincent, AAG, July 18, 1870, p. 93, Roll 1, M1002, USMA. For the "slop pail" incident, see Testimony of James W. Smith, ibid., p. 11. For the threats against Smith on his first night, see Testimony of Charles A. Wooden, ibid., p. 89, which is corroborated in Testimony of James Edward Shortelle, on pp. 80–81 of the same document; see also "West Point Gentlemen," *New York Herald,* undated clipping included in James W. Smith Files, M1002, USMA.

3. For more on Smith's experience at West Point and the dynamics of cadet and faculty resistance to his presence among them, see Rory McGovern, Makonen Campbell, and Louisa Koebrich, "'I Hope to Have Justice Done Me or I Can't Get Along Here': James Webster Smith and West Point," *Journal of Military History* 87.4 (October 2023): 964–1003.

4. This interpretation has its roots in Henry O. Flipper's memoir, which contains several statements throughout the book stating that the West Point staff and faculty were perfectly fair and unbiased in their dealings with him. In one fairly representative selection, Flipper wrote, "The officers of the institution have never, so far as I can say, shown any prejudice at all. They have treated me with uniform courtesy and impartiality." See Henry O. Flipper, *The Colored Cadet at West Point: Autobiography of Lieut. Henry O. Flipper, U.S.A., First Graduate of Color from the U.S. Military Academy* (New York: Homer Lee & Co., 1878), 122. For reasons that will be explored later in this essay, readers should not accept Flipper's statement at face value. Two books from the 1960s amplified this interpretation: see Thomas Fleming, *West Point: The Men and Times of the United States Military Academy* (New York: William Morrow, 1965), 213–31; and Stephen Ambrose, *Duty, Honor, Country: A History of West Point* (Baltimore: Johns Hopkins University Press, 1966), 231–37. These two books have had an outsized impact on subsequent histories of West Point and integration at West Point, which all too often repeat that same claim and cite these books as their evidence. For example, see William P. Vaughn, "West Point and the First Negro Cadet," *Military Affairs* 35.3 (October 1971): 100–102; Walter Scott Dillard, "The United States Military Academy, 1865–1900: The Uncertain Years" (PhD diss., University of Washington, 1972), ch. 7; and Tom Carhart, *Barricades: The First African American West Point Cadets and their Constant Fight for Survival* (Xlibris, 2020), chs. 3–4. The few exceptions are studies that focus more specifically on Johnson C. Whittaker's well-publicized assault and court-martial. In most cases Whittaker's assault pushes historians to conclude that the response of West Point's faculty and leadership was, at least in Whittaker's case, as problematic as the response of the wider corps of cadets. See Donald B. Connelly, *John M. Schofield and the Politics of Generalship* (Chapel Hill: University of North Carolina Press, 2006), ch. 11; and John F. Marszalek, *Assault at West Point: The Court-Martial of Johnson Whittaker* (New York: Collier Books, 1994). Surprisingly, however, some still defend the response even in Whittaker's case. See Carhart, *Barricades*, ch. 5.

5. For example, see Brian G. Shellum, *Black Cadet in a White Bastion: Charles Young at West Point* (Lincoln: University of Nebraska Press, 2006).

6. "Tabular Statement of Cadets Admitted into the United States Military Academy from its Origin till September 1, 1901," printed in *List of Cadets Admitted into the United States Military Academy, West Point, N.Y., From its Origin til September 1, 1901* (Washington, DC: Government Printing Office, 1902), 112–15.

7. Eric Foner, *Reconstruction: America's Unfinished Revolution, 1863–1877* (New York: Perennial Classics, 2002), xxiii

8. The postwar years brought a general decline in discipline among the corps of cadets as officers assigned to West Point infrequently enforced policies and regulations that appeared to them to be insignificant in light of their wartime experiences. Subpar superintendents serving between 1864 and 1871 only exacerbated this trend. See Theodore J. Crackel, *West Point: A Bicentennial History* (Lawrence,: University Press of

Kansas, 2002), 141–45; Lance Betros, *Carved from Granite: West Point since 1902* (College Station: Texas A&M University Press, 2012), 23–24; David J. Fitzpatrick, *Emory Upton: Misunderstood Reformer* (Norman: University of Oklahoma Press, 2017), 128–30; and Rory McGovern, *George W. Goethals and the Army: Change and Continuity in the Gilded Age and Progressive Era* (Lawrence: University Press of Kansas, 2019), 8.

9. "Grant vs. Smith," *New York Daily Tribune,* July 31, 1872.

10. James W. Smith to "Friend" [David A. Clark], June 29, 1870, included as an exhibit appended to Untitled Transcript of Court of Inquiry addressed to Thomas M. Vincent, AAG, James W. Smith Files, M1002, USMA.

11. Legrand W. Perce to Secretary of War, April 20, 1870, and Michael Howard and Merryman Howard to Secretary of War, April 30, 1870, Michael Howard Files, M1002, USMA; "The West Point Revels," *New York Sun,* May 26, 1870, M1002, USMA.

12. Solomon Hogue to Secretary of War, May 23–24, 1870; and James W. Smith and Israel Smith to Secretary of War, May 27–29, 1870, James W. Smith files, M1002, USMA.

13. Testimony of Henry M. Black, Untitled Transcript of Court of Inquiry addressed to Thomas M. Vincent, AAG, pp. 42–43, James W. Smith Files, Roll 1, M1002, USMA.

14. "The West Point Revels," *New York Sun,* May 26, 1870, Michael Howard Files, Roll 1, M1002, USMA.

15. J. W. Smith to "Friend and Benefactor," June 1, 1870, included as an exhibit appended to Untitled Transcript of Court of Inquiry addressed to Thomas M. Vincent, AAG, James W. Smith Files, Roll 1, M1002, USMA.

16. "The West Point Revels," *New York Sun,* May 26, 1870, Michael Howard Files, Roll 1, M1002, USMA.

17. See "Report of Difficulty between New Cadets Michael Howard and Rob't C. McChord, U.S.M.A.," June 17, 1870, Michael Howard Files, M1002, USMA. The report includes several attached files. Quotations are from Statement [of Michael Howard], June 6, 1870, and A. Clarke, "Statements of Cadets," both of which are appended to the report. See also Michael Howard to Adelbert Ames, June 9, 1870, ibid.

18. Testimony of James W. Smith and Henry M. Black, Untitled Transcript of Court of Inquiry addressed to Thomas M. Vincent, AAG, pp. 11, 44, James W. Smith Files, M1002, USMA.

19. James W. Smith to "My Kind Benefactor" [David A. Clark], June 11, 1870, included as an exhibit appended to Untitled Transcript of Court of Inquiry addressed to Thomas M. Vincent, AAG, James W. Smith Files, M1002, USMA.

20. Testimony of James W. Smith, Untitled Transcript of Court of Inquiry addressed to Thomas M. Vincent, AAG, pp. 21–25, James W. Smith Files, M1002, USMA. See also "West Point Gentlemen," *New York Herald,* July 7, 1870. The latter quote is from testimony of James Crane, Untitled Transcript of Court of Inquiry addressed to Thomas M. Vincent, AAG, pp. 75–76, James W. Smith Files, M1002, USMA.

21. See Testimony of Charles A. Wooden, Untitled Transcript of Court of Inquiry addressed to Thomas M. Vincent, AAG, p. 89, James W. Smith Files, M1002, USMA.

This is corroborated in Testimony of James Edward Shortelle, ibid., pp. 80–81. See also "West Point Gentlemen," *New York Herald,* undated clipping included in ibid.

22. Unpublished Memoir, p. 11, Eben Swift Files, USMA.

23. The incident is recounted in a June 29, 1870, letter from James W. Smith to David A. Clark, reprinted in "West Point Gentlemen," *New York Herald,* undated clipping included in James W. Smith Files, M1002, USMA. These particular quotations are unchanged in the copy of the original letter that Smith transcribed from memory and introduced as evidence in a court of inquiry—James W. Smith to "Friend" [David A. Clark], June 29, 1870, included as an exhibit appended to Untitled Transcript of Court of Inquiry addressed to Thomas M. Vincent, AAG, ibid. Although many cadets and the commandant himself flatly denied Smith's allegations, the first-class cadet serving as quartermaster of the summer encampment definitively corroborated them. See Testimony of Andrew H. Russell, ibid., pp. 93–95, 147. Russell's testimony is quite convincing.

24. James Fornance to Brother, July 17, 1870, James W. Smith Vertical File, USMA.

25. Testimony of Charles A. Wooden, Untitled Transcript of Court of Inquiry addressed to Thomas M. Vincent, AAG, p. 91, James W. Smith Files, M1002, USMA.

26. Peter S. Michie, *The Life and Letters of Emory Upton, Colonel of the Fourth Regiment of Artillery and Brevet Major-General, U.S. Army* (New York: D. Appleton and Co., 1885), 252; and Unpublished Memoir, pp. 12–13, 22–25, Eben Swift Files, USMA.

27. William Harding Carter to "Ma," November 5, 1871, Box 6, William Harding Carter Papers, US Army Heritage and Education Center, Carlisle, PA (hereafter referred to as USAHEC); Hugh Lenox Scott, *Some Memories of a Soldier* (New York: Century Co., 1928), 14. On hazing and indiscipline in the corps of cadets after the Civil War, see Ambrose, *Duty, Honor, Country,* 222–31; Crackel, *West Point: A Bicentennial History,* 141–45; Betros, *Carved from Granite,* 22–24; and Donald B. Connelly, "The Rocky Road to Reform: John M. Schofield at West Point, 1876–1881," in *West Point: Two Centuries and Beyond* (Abilene, TX: McWhiney Foundation Press, 2004), 175–78.

28. See "Report of Difficulty between New Cadets Michael Howard and Rob't C. Mc-Chord, U.S.M.A.," June 17, 1870, Michael Howard Files, M1002, USMA.

29. Court-Martial Case Files, Transcript of Second Day, October 21, 1870, James W. Smith Files, M1002, USMA.

30. James W. Smith to the editor of the *New National Era,* reprinted in Flipper, *Colored Cadet at West Point,* 297–99. The exact timing of this incident is unclear, but Smith introduces it in this letter in a way that makes it certain it happened during his first year. This incident does not appear in the archives, as Smith accepted a written apology from his assailant in lieu of asking the commandant to investigate and press charges: he had discerned the fact that investigations opened in response to his own complaints only caused more trouble for himself.

31. William Harding Carter diary, November 10, 1870, Box 6, William Harding Carter Papers, USAHEC. When accessed in the fall of 2021, USAHEC had not yet fully processed this somewhat recently donated collection; there is no guarantee that this

material will remain in Box 6 once the collection is processed. The same applies for subsequent citations from this collection.

32. Full transcripts from one court of inquiry and two courts-martial in 1870–71 are available in James W. Smith Files, M1002, USMA.

33. Unpublished memoir, p. 19, Eben Swift Files, USMA.

34. "The Colored Cadet Persecution," *New York Daily Tribune,* January 23, 1871; Court-Martial Case Files, Transcript of Third Day October 22, 1870, James W. Smith Files, M1002, USMA; Copy of Charge Specifications, undated [but clearly from January 1871 court-martial], ibid.

35. See unpublished memoir, p. 13, Eben Swift Files, USMA. Here Swift relates a narrowly avoided fight between a plebe and an upperclassman. He was relieved the would-be pugilists thought the better of it. He recalled, "This was a critical moment for me in my career. If the fight had come off and I had told the truth at any investigation it would have created a prejudice against me, which might have seriously interfered in my future life, however illogical that might be."

36. Examples derived from Testimony of Henry S. Taber, Untitled Transcript of Court of Inquiry addressed to Thomas M. Vincent, AAG, p. 65; Testimony of George R. Smith, ibid., p. 114; and Testimony of Quincy O. Gillmore, ibid., p. 50, all in James W. Smith Files, M1002, USMA.

37. Copy of Charge Specifications, undated [but clearly from January 1871 court-martial], James W. Smith Files, M1002, USMA; Court-Martial Case Files, Transcript of Third Day October 22, 1870, ibid.; and Emory Upton to Capt. Boynton, Adjutant USMA, April 1, 1871, ibid.

38. J. Mott to Secretary of War, January 23, 1871, James W. Smith Files, M1002, USMA. See also "The Colored Cadet," *New York Daily Tribune,* January 10, 1871.

39. Transcript from the morning of July 21, 1870, Untitled Transcript of Court of Inquiry addressed to Thomas M. Vincent, AAG, James W. Smith Files, M1002, USMA; William Belknap, Memorandum, August 10, 1870, M1002, USMA; The Colored Cadet," *New York Daily Tribune,* January 10, 1871.

40. Unsigned Memorandum from Secretary of War, William Belknap, Undated, James W. Smith Files, M1002, USMA.

41. J. Mott to Secretary of War, January 23, 1871, James W. Smith Files, M1002, USMA; General Court-Martial Orders No. 6, June 18, 1871, ibid.

42. Both quotations from William Harding Carter to "Ma," March 19, 1871, Box 6, William Harding Carter Papers, USAHEC.

43. This is the central argument in McGovern, Campbell, and Koebrich, "I Hope to Have Justice Done Me."

44. Flipper, *Colored Cadet at West Point,* 37, 106–7.

45. George W. Goethals to "Friend Hank," May 19, 1877, George W. Goethals Vertical File, USMA.

46. "Statement Showing the Number of Colored Persons Appointed to the U.S. Military Academy," October 21, 1886, USMA. Natural and experimental philosophy felled

Smith in June 1874. The Academic Board declared Henry Alonzo Napier deficient in mathematics and French in June 1872. Thomas Van Rensselaer Gibbs was likewise declared deficient in mathematics in January 1873. John Washington Williams failed French in January 1874. And Johnson Chestnut Whittaker—admitted in 1876—was declared deficient in natural and experimental philosophy in 1880 and discharged after his lengthy court-martial concluded in 1882.

47. Testimony of Edward Hardin, Untitled Transcript of Court of Inquiry Addressed to Thomas M. Vincent, AAG, July 18, 1870, p. 111, Roll 1, M1002, USMA.

48. James W. Smith to the editor of the *New National Era,* August 19, 1874, reprinted in Flipper, *Colored Cadet at West Point,* 299–304.

49. Flipper, *Colored Cadet at West Point,* 120.

50. See Testimony of Andrew H. Russell, Untitled Transcript of Court of Inquiry addressed to Thomas M. Vincent, AAG, pp. 93–95, 146–47; Testimony of Charles Woodruff, pp. 124–27; and Testimony of Henry M. Black, pp. 42–46, all in James W. Smith Files, M1002, USMA.

51. Registry of Delinquencies, Classes of 1873 and 1874, p. 257, USMA.

52. Unpublished memoir, pp. 48–49, Eben Swift Files, USMA. The doggerel was titled "Nigger Jim." Lyrics make it clear that the doggerel was written at some point after Smith's failed academic examination, so it is unclear if it was ever performed in his presence. But we know it was performed often enough before Eben Swift graduated in 1876 that he remembered the tune and lyrics precisely when he wrote his memoir in 1926, which means that even if it was not performed in front of James W. Smith, it may have been performed in front of Henry O. Flipper. Swift recalled the lyrics:

I'm de noted Cullerd Ca-det
And from Dixie land I came,
Where I used to hoe de cotton
And de cane, all de day
All de day (basso profundo)
x x x x x x x
Old Davy's done it for you now
Dis Nigga' he has found
And you wont see any more of this yere chile
Dis yere chile (*basso profundo*)
But I'll write a book on West Point
And for Congress I will run
And I'll engineer a bill to hang old Lyle
G— D— old Lyle (*basso profundo*)

In a previous song recorded on page 47, "Davy" refers to Lt. David S. Lyle, who worked for Prof. Peter Michie in the Natural and Experimental Philosophy Department.

53. Flipper, *Colored Cadet at West Point,* 37.

54. George W. Goethals to "Friend Hank," April 28 and May 19, 1877, George W. Goethals Files, USMA.

55. Excerpt of letter from Johnson C. Whittaker to Henry O. Flipper, reprinted in Flipper, *Colored Cadet at West Point,* 281.

56. Flipper, *Colored Cadet at West Point,* 122.

57. Crackel, *West Point,* 294.

58. Ambrose, *Duty, Honor, Country,* 202–3.

59. George L. Andrews, "West Point and the Colored Cadets" *International Review* 9 (November 1880): 477–98, quotes on 484.

60. Peter S. Michie, "Caste at West Point," *North American Review* 130.238 (June 1880): 611.

61. See "Report of Difficulty between New Cadets Michael Howard and Rob't C. McChord, U.S.M.A.," June 17, 1870, Michael Howard Files, M1002, USMA.

62. Testimony of Henry M. Black, Untitled Transcript of Court of Inquiry addressed to Thomas M. Vincent, AAG, p. 44, James W. Smith Files, M1002, USMA. The absence of archival records of an investigation is very unusual. Records in the USMA Library Special Collections and Archives are replete with investigations that range across a broad spectrum of relative significance.

63. James W. Smith to the editor of the *New National Era,* August 19, 1874, reprinted in Flipper, *Colored Cadet at West Point,* 299–301.

64. James W. Smith to the editor of the *New National Era,* August 13, 1874, reprinted in Flipper, *Colored Cadet at West Point,* 297–99, quote 299.

65. Flipper included only "'Bentz the bugler,' the tailor, the barber, commissary clerk, the policeman who scrubbed out my room and brought around the mail, the treasurer's clerk, cadets occasionally, and others" on that list. See Flipper, *Colored Cadet at West Point,* 106–7.

66. "Statement Showing the Number of Colored Persons Appointed to the U.S. Military Academy," October 21, 1886, USMA. The five were James Webster Smith, Henry Alonzo Napier, Thomas Van Rensselaer Gibbs, John Washington Williams, and Johnson Chestnut Whittaker. Whittaker's failed exam was in 1880, but he had been admitted in 1876. Of these, Smith and Whittaker's dismissals were for failures in natural and experimental philosophy; Williams's was for French; Napier's was for math and French; and Gibbs's was for math.

67. Testimony of James W. Smith, Untitled Transcript of Court of Inquiry addressed to Thomas M. Vincent, AAG, pp. 32–33, James W. Smith Files, M1002, USMA; and Testimony of Emory Upton, ibid., pp. 149–50. For Upton's abolitionism, see Fitzpatrick, *Emory Upton,* 12–24; and Michie, *Life and Letters of Emory Upton,* 2–7.

68. Michie, *Life and Letters of Emory Upton,* 252.

69. Emory Upton to Capt. Boynton, Adjutant USMA, April 1, 1871, James W. Smith Files, M1002, USMA.

70. Court-Martial Case Files, Transcript of Third Day October 22, 1870, James W. Smith Files, M1002, USMA; "Copy of Charge Specifications," undated [January

1871 Court-Martial], ibid.; James W. Smith to the editor of the *New National Era,* August 19, 1874, reprinted in Flipper, *Colored Cadet at West Point,* 302–3.

71. J. Mott to Secretary of War, January 23, 1871, James W. Smith Files, M1002, USMA.

72. Flipper, *Colored Cadet at West Point,* 37.

73. Unpublished memoir, pp. 47–49, Eben Swift Files, USMA. The doggerel itself appears on pp. 48–49. That "Davy" refers to Lyle becomes apparent when analyzing the doggerel alongside other song lyrics printed on p. 47.

"THE ARMY WOULD SOON GROW ACCUSTOMED TO IT"

The Army Reacts to Integration at West Point

Jonathan D. Bratten

In June 1870, Brig. Gen. Alfred Terry sent an impassioned letter to William T. Sherman, commanding general of the US Army. As military governor of Reconstruction-era Georgia, he begged Sherman's advice on how to deal with the question of race as he navigated the pitfalls of military governance. The local freed people had come to him, Terry explained, and asked him to enact a new policy that would ensure multiracial juries, as they were otherwise unable to get a fair trial. Terry admitted that they were in the right and stated that he hoped they would allow the state legislature to pass this as a permanent law rather than resort to temporary military policy. He admitted to Sherman that he personally doubted that it would pass. Frustrated, he acknowledged, "I confess that I am a little sick of being appealed to to remedy every fault in the laws of this State."[1] Officers across the US Army were trying to navigate a path through the new world of racial equality before the law amid Reconstruction, and many were finding it difficult. The issue of having a fair trial would soon be the question not only of freed people in the south but also of those wearing cadet gray on the Hudson.

When the US Military Academy began admitting African American cadets in 1870, it captured the attention of the entire nation. Nowhere was this topic more intensely watched than in the ranks of the professional officer corps. The army of the 1870s and 1880s was a highly stratified, conservative, and traditional institution. Scattered across multiple posts on the western frontier, the Reconstruction South, and coastal fortifications on the East Coast, officers used professional periodicals such as the *Army and Navy Journal* not only as sources of news and information but also as vehicles for opinion and debate. Across the pages of these periodicals, officers and veterans debated the issues of the day, from the Franco-Prussian War to Reconstruction to the issue of race at West Point.

The journal began in 1863 as the *Army and Navy Journal and Gazette of the Regular and Volunteer Forces,* founded by New York City brothers William and Francis Church as a "a weekly newspaper, devoted to the interests of the Army and Navy, and to the dissemination of correct military information."[2] Both men had worked for the *New York Times* before the war, but William left the *Times* during the War of the Rebellion to take a volunteer commission in the US Army in 1862. He was wounded shortly after entering the army during the Peninsula Campaign. He left the army in 1863 to begin production of the *Journal* with his brother, Francis, who had covered the war for the *Times* before moving on to the *Sun,* where he would later pen his most famous editorial, "Yes, Virginia, there is a Santa Claus" in 1897.[3]

The *Journal* was something entirely novel in the world of American journalism—it was geared wholly toward the military establishment. Within the historiography of professional publications in the US military, however, the *Army and Navy Journal* does not hold the highest rank. As Dr. Richard Stewart wrote of the *Journal,* it differed from subsequent professional military journals, "but along with its social and other items about service personnel it carried articles, correspondence, and news of interest to military people that helped bind its readers together in a common professional fraternity."[4] However, in its own way, the *Journal* actively contributed to the movement toward a professional officer corps in the army. The army itself was in the midst of professionalizing in this era, following lessons learned from the War of the Rebellion. Although the *Journal* was not started or managed by career officers, who would have considered themselves military professionals, the debates on its pages helped move the army from a constabulary force in the mid- to late nineteenth century to a professional force in the early twentieth century. Within this context, the *Journal* can be considered a professional publication. The *Journal* continued publication into the twenty-first century as the *Armed Forces Journal.*[5]

The *Army and Navy Journal,* then, was not managed by graduates of West Point. Nor were they professional soldiers. William Church (who would remain the managing editor until 1917 and would also be a cofounder of the National Rifle Association, established to help National Guardsmen become better marksmen) was an officer of volunteers during the war.[6] He owed no loyalty to the US Military Academy from any type of affiliation as a cadet or even as an officer in the regular army. And yet the pages of the *Journal* in the

1870s and 1880s ring as hearty with endorsements of West Point as if William Church had been the proudest of graduates. On race, Church was adamantly opposed to slavery, was proemancipation when the time came, and advocated that the army stop using pejorative terms for African Americans.[7] At the same time, he was not a prewar abolitionist and was reluctant to integrate African Americans into social norms, positions he held in common with many white northerners in that era.

The views of the other editors and authors writing for the *Journal* are another matter entirely. While the *Journal* stated, "It is necessary that the name of the writer should, in all cases, accompany his communications, not for publication, but as a guarantee of good faith," the use of published pseudonyms in the *Journal* was common. Indeed, it was common practice for many journalists of that era to use pseudonyms. Sam Wilkeson of the *New York Tribune*—whose son Bayard was killed during the war while commanding a battery at Gettysburg on Barlow's Knoll—wrote that he almost always required his authors to use a pseudonym. "The anonymous greatly favors freedom and boldness in newspaper correspondence," he directed. "Besides the responsibility it fastens on a correspondent, the signature inevitably detracts from the powerful impersonality of a journal."[8] So while the *Journal* was in line with other publications of the time in its use of pseudonyms, this practice obscures the identities and ranks of many of those who authored letters or articles published in in its pages surrounding Black cadets at West Point. That said, we can infer from the *Journal*'s guidelines of accepting pieces primarily from those serving that those writing were either active members or veterans, and so the letters, articles, and editorials therein offer a good indicator of widespread army opinions. The *Journal* also serves as our only window into the army at large in this era. Branch journals would not begin to circulate until the late 1880s. Two professional journals, *United Service* and the *Journal of the Military Service Institution of the United States,* began in the late 1870s and early 1880s. However, a survey of both found no mention of Black cadets at West Point. Branch-specific journals did not appear until even later. Thus, with its editorials, letters to the editor, press excerpts concerning the army, and publication of official correspondence, the *Army and Navy Journal* is our best window into the army's perspective during the 1870s and 1880s. This resource is by no means perfect. Its anonymity unfortunately means that we cannot distinguish the differences of opinion between officers and enlisted soldiers.

The admission of West Point's first Black cadet in June 1870 was not, according to the *Journal,* an earth-shaking event. The authors and editors, and most of the US Army itself, were more riveted by the ongoing Franco-Prussian War (1870–71), which began just weeks after Cadet James Webster Smith passed his entrance exams and was admitted to West Point. It should hardly be surprising that officers of the day could not help but to compare the new conflict with their own experiences in the most recent war, only five years in the past.

The first mention of Cadet Smith did not cast him in a promising light. In July 1870 the *Journal* mentioned that there was a "colored cadet" at West Point—noting that a court of inquiry had investigated the allegations that the cadet had been hazed by his fellows. The finding was negative. The court had not found in favor of Smith; rather, it found that it was "'devilling' of the sharpest kind" which the "whitest youth is obliged to undergo an ordeal of."[9] The court reprimanded Smith and one white cadet whose behavior had been too egregious to be ignored.

In striking contrast, that same issue ran a letter from Brevet Maj. Gen. Oliver O. Howard that had originally been published in the *New York Tribune*. Howard—then commissioner of the Freedmen's Bureau—published a letter he had received from Cadet Smith that detailed the cadet's difficult time at the academy: hazing, isolation, and embarrassment. Howard's own public letter was a challenge to the academy: "If West Point has not power enough to protect such a young man as Cadet Smith—quick, able, honest, noble-spirited as he is—then West Point will have a hard struggle against the returning tide of feeling that will break in from the people. I am a graduate of West Point, and am proud of her sons who have been true to the country and true to humanity, but I am greatly ashamed when cadets dishonor us by a mean prejudice, that ought long ago to have been smothered."[10] Howard would be one of the sole general officers speaking out in defense of Black cadets.

It was not until September that some in the army took to the pages of the *Journal* to pay attention to Smith. In the "Correspondence" section of the September 10, 1870, edition, sandwiched between letters about baseball and army cooking, there was a sign that all was not going as expected at the US Military Academy at West Point. In a letter from a reader signed only "M.B.S.," the author bemoaned the fact that West Point was surely declining because he believed the habit of "devilling plebes" (hazing) was no longer in vogue. He elaborated as to one of the causes: "If there lingers one trace of it now, since a negro cadet has appeared on the arena, devilling must be considered as numbered with the things that were, and the best lesson ever taught at the academy is lost."[11] Without even the mention of his name, James Webster Smith was being labelled as the cause for lowered standards at the academy.

Smith was eventually fully named in that same issue, but that was several pages on and following the descriptions of the capture of Napoleon III at the Battle of Sedan and the rise of the new French Republic. Here, in an unnamed editorial, Smith was introduced via a vignette from the *New York Tribune* concerning an incident in which Smith got into a physical altercation with another cadet over access to well water. The *Tribune* noted that Smith's action of self-defense was looked on commendably by the other cadets and the officers, which seemed to them to give hope of "fair official treatment, and redress in all injustice, of the colored cadet."[12] The impersonal use of the phrase "the colored cadet" would describe not only James Webster Smith but also Johnson Chestnut Whittaker, Henry Ossian Flipper, or any other African Americans who later attempted to penetrate the academy's social norms in the nineteenth century.

Despite the *Journal's* assurances, the academy did not commend Smith for the well water fight and instead pursued legal action. Smith's altercation resulted in three weeks of incarceration and the first of what would be two courts-martial for the cadet in his first year.[13] The *Journal's* assurance that Black cadets would receive "fair official treatment" ("This has, of course, been thoroughly understood all along by those who know anything of the Military Academy") began to sound weak as the court-martial boards continued.[14] The *Journal* would run portions or all of the court-martial proceedings, the editors often commenting that care was needed to be impartial regarding "the negro question" because of how public perception might damage the academy's image.[15]

This constant care for the academy's image reveals more about the editors than they probably realized. Their assurances that Smith was receiving fair and impartial treatment show that they cared far more about the academy's reputation than about Smith's guilt or innocence—or even well-being. Indeed, their insistence on impartiality may even point to their belief that Smith was guilty before the court-martial had even concluded. This overprotective trend toward appearing objective to the point of obsession revealed their predisposed bias, which continued as long as Black cadets were attending West Point in the nineteenth century.

By January 1871, Smith was on his second court-martial. The *Journal* once again opined that West Point was the model of institutional fairness. After all, "in every college in the country the negro question looms up, to the dismay of many faculties and students"; in that regard, they claimed that West Point would stand out by removing "every shadow of prejudice."[16] How that could be managed when all the officers and witnesses were white the *Journal* did not mention. But although the *Journal* covered the trial, its editors did not deem it to be the biggest news out of the academy worth noting. "Political papers are making a great noise over the trial of Cadet Smith and the 'oppression' which he suffers," scoffed the editors on January 14, 1871. "But the jealousies, quarrels, and pranks of the fourth class ([plebes, or freshmen] are of little moment."[17] The great and shattering news out of the hallowed halls of the United States Military Academy came from the actions of several of the first-year (firstie, or senior) cadets who had decided to take justice into their own hands and deal with cadet infractions in their own way.

In an incident that would have been described merely as "collegiate" anywhere else, three plebes had managed to get themselves into trouble by sneaking off academy grounds with a bottle of whiskey at night. This was not uncommon, but the first-class cadets felt that in this case, the "honor of the corps" was at stake because the wrongdoers had lied to them when questioned about it. The first offense was mild, but misrepresenting it after the fact and breaking the cadet honor code was deemed a far worse offense. And so they resolved on a stern punishment. In the dead of night, the three plebes were rousted out of their beds by the first-class cadets, made to put on civilian clothes, brought to one of the academy's gates near the site of the American Revolution–era Fort Putnam, and banished. The offenses of the first-year cadets were made all the worse when the news came out that among them was Frederick Dent Grant, President Ulysses S. Grant's son.

In the world of West Point, this breakdown in the official chain of command was a "very grave matter of usurpation of power by the first class." How grave? As written by the superintendent, Col. Thomas Pitcher, in his Special Orders No. 3, "No words can express too forcibly the disapproval by the superintendent of the assumption of powers by cadets, until now unknown to those who organized, and for 68 years have governed, the Military Academy."[18] If this seems like hyperbole in light of the actions of some former West Pointers who between the years 1861 and 1865 attempted to destroy the Union, bear in mind that Pitcher's consternation is precisely *because* of this. After the Civil War, West Point faced a grave reckoning. Long the target of suspicion before the war by many who thought the academy was producing an elitist, upperclass military aristocracy, this rancor was doubled after a war in which so many graduates of West Point fought for the Confederacy. Many critics were again calling for West Point's demise.

Incredibly, the man coming to West Point's defense in this matter in Congress was Benjamin Butler, who as an officer of volunteers in the War of the Rebellion had long deplored what he saw as a corrupting influence amongst academy graduates. "Now gentleman," he told Congress, "we have just forgiven more than five hundred West Point officers who had broken their oaths, and were guilty of treason and murder, and have removed their disabilities. Shall we now refuse to be merciful to these boys?" Butler had another pointed jab, perhaps also aimed at the academy's other public relations issue, that of the court-martial of Cadet Smith. Butler reminded his listeners that on "account of the jealousy of West Point officers," he had not been allowed to enter Richmond in 1865; still, he was grateful that "the brave Twenty-fifth corps of colored troops, every regiment of which had been recruited under his direction . . . was one of the first to go into Richmond."[19] Few who were listening could fail to see the veiled inference from Butler: West Pointers would not give Smith a fair trial. Butler himself had long been a proponent of Black cadets at West Point—he had advocated for James Gregory to enter the academy in 1868, relenting only after it became clear that Gregory was younger than the statutory minimum age for West Point cadets. He observed Smith's court-martial with considerable concern.[20]

This was the backdrop for Smith's first court-martial. The *Journal* deplored that this affair of "boys will be boys" had made its way to Congress, where the academy was "to be made the football of spiteful politicians."[21] With the *Journal* representing a pro–West Point perspective, their reporting took on a very defensive tone on anything that might have seemed critical of the academy. Writing on January 21, 1871, an unnamed *Journal* correspondent took civilian reporters to task for saying that Cadet Smith had been ostracized: "This is so absurd that it scarcely needs a total denial of its having any foundation whatever. . . . To say the contrary is to do rank injustice to the honorable and Christian gentlemen, his instructors and superior officers."[22] Smith's second court-martial sentenced him to be dismissed. The judge advocate general of the army persuaded the secretary of war and President Grant to suspend the sentence, and in April 1871 Smith was instead ordered to repeat his plebe year.

By February, the *Journal* took its hardest and most defined line not only on the Smith case but also on the case of Black officers in the US Army. The timing of this piece was auspicious, coming as it did after Smith's verdict of dismissal and as the War Department considered the sentence, both situations about which the *Journal's* readers were well aware. In an editorial published on February 25, 1871, the unnamed author decried the political forces in play at West Point, with their "hasty, ill-advised, and vicious legislation." The author wove a clear picture of all that the army had done to aid the "slow advance of civilization westward" and how the military was the bedrock of society. Congress, said the author, was conducting experiments with the army and with West Point when it "commenced sending colored cadets to be educated for commissions in the army. The experiment in its nature is a dangerous one." Over the next page and a half, the author listed the reasons for this danger. It would put the "the Anglo-Saxon and African on equality" and force "the social problem in the wrong place." There would be a breakdown of army readiness: "Sift out the 30,000 troops now in our Army, and from them all not one company could be organized that would consent to be commanded by a colored officer." Lastly, if there were Black officers—and by extension Black wives of these officers—the combination would "destroy garrison life as it is made by these [white wives of officers, who were] self-sacrificing and true hearted women." This, the editorial claimed, would "effectually destroy the Army."[23] With that, the *Journal* gave up any indication that it was impartial in the case of Cadet Smith.

In the matter of Black naval midshipmen, the *Journal* took a different tone. When African American midshipman James Conyers was admitted to the US Naval Academy in 1872, he began to experience the same hazing and discrimination that Cadet Smith experienced at West Point. The *Journal* had a radically different reaction to this than they had to Smith's ordeals. On October 19, 1872, the *Journal* admonished the Naval Academy's midshipmen that "abusing this negro cadet on account of the accident of his color" was "making a foolish mistake. . . . The colored population has surely its right to seek representation in the Army and Navy, if it chooses to do so." Since the author was not named, we cannot know if this was the same author who had come down so negatively on Smith's case. But as if to qualify this abrupt leap toward egalitarianism, the *Journal* concluded that the American principle was that "not all men are equal, but that every man has the right to be the equal of every other man if he can."[24] Like Smith, Conyers would also get into an altercation with fellow midshipmen. Unlike the US Military Academy, the Naval Academy expelled the white midshipmen involved in the incident. On this, the *Journal* opined: "As an impotent protest against the spirit of national law, such conduct was folly; as an insolent expression of superiority to the solitary representative of a lately oppressed race, it was unmanly."[25] The disparity between the army and the navy's treatment of prospective Black officers is clearly indicated—as is the disparity in the way that the *Journal* reported these stories.

When Cadet James Webster Smith was dismissed from West Point in 1874, the *Journal* barely bothered to give him notice. He received one paragraph noting that he had left of his own volition—which was untrue—and that the "disease of 'civil rights,' as applied to the selection of cadets, left to run its course, has wrought its own cure." Henry Ossian Flipper's arrival at the academy was first given notice here, although the *Journal* only mentioned that "another colored cadet" had arrived at the academy who was sure to do better than Smith.[26]

In all of Smith's tenure at West Point, the only person to speak up for him in the pages of the *Journal* was Major General Howard. Howard was a true abolitionist and was the head of the Freedmen's Bureau after the war, so it is not a stretch to say that he was in the minority in his defense of Smith.

The position of the writer quoted above was probably a true representation of general opinion in the army. Of course, a small group of regular officers and volunteer veterans likely would have been horrified at Smith's treatment—men like Maj. Rufus Saxton, who had worked with US Colored Troops during the war and for the Freedman's Bureau during Reconstruction. A West Pointer himself, Saxton was a pointed critic of the academy, writing to Benjamin Butler in 1870: "If West Point shall fail in the future to represent those advanced opinions upon the equality of all men before the law . . . then it will be unworthy of the support of a government which respects the rights of all governed." Butler, too, was an outspoken advocate for Black cadets. His son Bennie attended West Point with Flipper and apparently absorbed his father's progressive views. Flipper commented later that "among the three hundred cadets I hadn't a better friend than the son of the Massachusetts statesman."[27] However, Howard, Saxton, and Butler were a minority of radical officers. The more conservative majority would have sided with the academy, as evidenced by their complete silence on the issue of Smith's departure.

A letter from a correspondent in June 1875 noted that "Flipper," as the *Journal* called him, was the only Black cadet at West Point and was doing better than James Smith had done. He "minds his own business, and he does not intrude his company upon the other cadets."[28] Clearly, the author considered that Black cadets should banish themselves for the good of the group. Flipper made few appearances in the *Journal* during his time at West Point, being mentioned again only in relation to Cadet Charles A. Minnie, whose time at the academy was brief, having failed several examinations. Flipper's graduation and subsequent commissioning in June 1877 merited only a few lines in the *Journal,* which ran an excerpt from the *New York Times* extolling how Flipper endured ostracism "quietly and bravely" for four years.[29] The principal difference between James Smith and Henry Flipper, in the eyes of the *Journal,* seems to have been that whereas Flipper bore his sufferings quietly, Smith had fought back. Again, the contrast between the *Journal's* commentary on Smith at West Point and Conyers at the Naval Academy is striking and may speak more to the editors' army backgrounds than anything else.

Henry O. Flipper's tenure at West Point had an interesting postscript in the pages of the *Journal,* even before the eventual court-martial that would see him dismissed from the army. Having just graduated a few months earlier, 2nd Lt. Flipper of the Tenth US Cavalry was rumored to be leaving the US to take the position of general-in-chief of the Liberian Army in October 1877. Flipper wrote to several newspapers to dispel this rumor: "I am not at all disposed to flee from one shadow to grasp at another—from the supposed error of Hayes' Southern policy [i.e., ending Reconstruction] to the prospective glory of commanding Liberia's army."[30] Flipper was letting it be known that while he might have been quiet as a cadet, he did not intend to be so now. This became very apparent when he released a memoir of his time at the academy in 1878, called *The Colored Cadet at West Point*. The *Journal* labeled it "an interesting account."[31]

The *Journal* never pretended that the Black cadets at West Point had an easy time of it. The difference in their coverage of them largely seems to have been based on whether they were submissive or if they pushed back against the prevailing system. For example, Cadet Minnie was greeted with welcoming words when he first began at West Point, with the *Journal* noting that he was clearly there on account of his merit. However, by 1878 the *Journal* had changed its perspective when Minnie publicly complained of his treatment after he left the academy. To oppose Minnie's accusations of unfairness, the *Journal* ran a letter from Richard T. Greener, the first Black graduate of Harvard, who noted that he trusted West Point's fairness and that the Black cadet he had tutored was doing well there.[32] The *Journal* used Greener's letter to show that success or failure at West Point was entirely dependent on merit rather than race. Ironically, the cadet Greener was speaking of was Johnson C. Whittaker, who would later become the most controversial cadet covered in the pages of the *Journal*.

In 1877, as President Rutherford Hayes was beginning to plan an end to the US Army's administration of the southern states, Cadet Whittaker entered West Point with little notice from the *Journal*. He would go unremarked for three years. This in itself is remarkable, given the frequency with which the *Journal* discussed Smith. It could possibly indicate the *Journal*'s editors' grudging, and fleeting, acceptance of Black cadets at West Point. Not until April 10, 1880, when the headline "A Strange Affair at West Point" greeted readers, did Whittaker enter the public view for the wider army. The *Journal* devoted a column and a half on reporting the incident of Whittaker's assault. It first provided Whittaker's statement and claim that he had been assaulted. The editor then followed that with a report from the *New York Herald*, which wrote "that the whole [Whittaker] affair is simply a bold, reckless, and yet puerile attempt at imposition—that Whittaker wounded himself, and that the commanding officers so believe."[33] The editors made sure to highlight the fact that Whittaker had done poorly on exams the previous year.

A week later the *Journal* ran a full account of the proceedings of Whittaker's first court of inquiry, taking up the better part of three pages. This was the most coverage any Black cadet at West Point had ever received, and the *Journal* would continue to cover the inquiries and investigations in detail.[34]

The case had stirred public opinion around West Point and the army, and the negative publicity had to be countered. "It is an extraordinary fact that the public interest of the country has been far more aroused by the affair of Cadet Whittaker," remarked the *Journal*, "than by the question of an entire reorganization of the Army or by a midwinter campaign against the Indians, in which a score of officers and men are wounded or slain." Following the three pages of coverage was a page-long editorial, in which the editor remarked that he was of the opinion that Whittaker was lying and that "*character* will here, as elsewhere, assert its worth and dignity."[35] This, in turn, was followed by a letter to the editor that theorized that the Whittaker case was merely another scheme from anti–West Point partisans to shutter the academy for good.[36] For the editors of the *Journal,* the Whittaker case was not about a cadet seeking justice but about the army seeking vindication of West Point.

One week later, on April 24, a *Journal* editor led the issue with an impassioned defense—not of Cadet Whittaker but of West Point. "Execrated from the pulpit, denounced in Congress, abused roundly by the press, and even losing the confidence of its friends," he wrote, "the Academy seems to be shut out from all sympathy of our countrymen." The academy—and, by proxy, the army—was receiving strong public condemnation for how quickly the investigation had turned against Whittaker. The *Journal,* as the unofficial voice of the army, then turned against the public: "The onus of the social difficulty, of which so much is made, rest[s], not upon the Academy, but upon the community in which negroes everywhere find themselves set apart in social intercourse." Black men had all the civil rights they needed to be successful at the academy, the editor opined. Yet, he wondered, "Since when has it been the custom in this or any other country to identify civil right with social privilege?"[37] In other words, this author argued that it was proper that Black cadets suffered ostracism from their peers and racism from within and without the academy and that nothing should be done to alleviate their plight. In this, the writer tapped into a common theme. Contemporary white officers angrily differentiated between equality before the law and social equality.

With every passing week of the Whittaker trial, there were fresh public attacks against the academy—attacks that the *Journal* took upon itself to vociferously defend against. One editor called all attacks on West Point political, taking special aim at the *National Republican,* which had said that West Point had lost the faith that the nation put in it. Retired Maj. Gen. Henry W. Slocum, a Union corps commander in the War of the Rebellion, wrote to express support for his alma mater, West Point. The general-turned-Democratic politician warned that this much interference from outside the academy was making it "evident that that the liberties of this country are in great danger."[38] The *Journal* ran an excerpt of a congressional debate in which one senator recommended that the president nominate two Black cadets every year, another recommended West Point be abolished, and yet another recommended that the experiment of Black cadets at West Point be ended.[39] The *Journal* even saw fit to run the highly critical account penned by the Reverend Henry Ward Beecher calling for greater transparency at West Point.[40] All of this was based on debate surrounding the Whittaker case. It was clear that both the *Journal* and the army community at large were irate at the sustained attacks against an institution considered to be the beating heart of the regular army. The angry rhetoric matches that of the two academy professors quoted in Rory McGovern's essay in this volume, who made public statements blaming society at large for creating a problem at West Point by forcing Black cadets on the academy.[41]

However, not all graduates of West Point closed ranks to defend the academy. One unnamed letter-writer posited that Whittaker's assault was due to the poor discipline at the academy. The trouble, the author related, began when "administration of this military school was . . . taken from the control of the Engineer Corps." The trouble continued when the position of superintendent became that of a major general and the posting came as a reward for good service rather than being assigned to the best-qualified person for the job. West Point, claimed the author, was letting its standards of discipline slip and was not holding true to its guiding principles. Discipline, noted the author, was the problem, not any racial bias at the academy. Replying to this relatively mild accusation from one of its own, the West Point apologists at the *Journal* simply stated, *"Et tu, Brute!"*[42]

By June 1880, the editors of the *Journal* had made up their minds concerning Whittaker: "If Cadet Whittaker were a White boy his case would soon be settled, for the evidence is positive and convincing that he wrote the letter of warning and perpetrated the 'outrage' on himself." The editors decreed that the only reason the case was ongoing was because of politics and public opinion.[43] At no time did the editors stop to engage with the idea that the army and West Point both depended on politics and public opinion for funding, for manpower, or, indeed, that the army was controlled by civilians. The closest they got was when they ran a letter from an anonymous author who wrote that since West Point closely represented American society, then should the institution not "reflect the sentiments and teaching of the plain people?"[44] As Whittaker entered a forced leave of absence in late 1880, the *Journal* admitted, via a "West Point newspaper correspondent," that there was no one "outside the Military Academy that believes Whittaker's injuries were self-inflicted."[45] This was, of course, a misstatement, as the *Journal* had demonstrated that the broader army community was perfectly willing to believe West Point's version of the story. Despite this, the editors of the *Journal* continued to ridicule Whittaker and the academy's detractors.

The Whittaker case had drawn more public attention to the army and to race in the military than any other single incident in the post-Reconstruction United States. So much so that when Gen. William T. Sherman gave his annual report on the US Army in November 1880, he was forced to address the Whittaker case. The *Journal* ran his remarks in near entirety. Sherman defended West Point, using much the same language that could have been found in the pages of the *Journal* since 1870. While he admitted that there was racial prejudice at West Point, he stated that "there is no more such prejudice at West Point than in the country at large, and the practice of social equality at West Point is in advance of the rest of the country."[46] Giving preferential treatment to a Black cadet, he said, would be in violation of the Fourteenth Amendment.

In relation to this amendment, Sherman counterattacked the academy's—and, by proxy, the army's—critics, by levelling this shocking challenge: "I desire to state that in my judgment, the requirement that all the enlisted men of the 9th and 10th Cavalry and of the 24th and 25th Infantry shall be colored men, while the officers are white, is not consistent with the amendment of the Constitution referred to. All men should be enlisted who are qualified and assigned to regiments regardless of color or previous condition. Such has been the law and usage in the Navy for years, and the Army would soon grow accustomed to it."[47]

The general-in-chief of the US Army was calling for its desegregation a full sixty-eight years before President Harry Truman ordered it in 1948. This was a bombshell of significant proportion, buried inside his annual report. How much Sherman really meant his shocking suggestion is open for question.[48] Personally, he remained committed to segregation. Writing to a friend in 1877, Sherman had written on the question of race in the army, "I honestly think the White race the best for this."[49] In a December 1880 interview that ran in the *Washington Post,* he remarked concerning a question about John Schofield's handling of Black cadets at the academy that "we do not keep up West Point to equalize the negro with the white man, but to make soldiers." He went on to say that the question of racial equality was not a political one, but one of social acceptance: "The negro will not be received in society and be invited to dinner to your house and mine until it becomes the fashion."[50] However, his statement on desegregation centered not on race but rather on constitutionality. Sherman saw the army's policy on segregation as a violation of the

Fourteenth Amendment. As an officer in the US Army who not only swore an oath to protect the Constitution but who had also fought to defend it, this contradiction did not sit right with Sherman. His position aligned with his continuous adherence to the Constitution, even if it differed from his personal beliefs. Regardless of his intent in making this statement, this was a striking divergence from conventional discussions around race in the US Army and must absolutely be taken into consideration when discussing the army's reaction to race and West Point.

In November 1880, the *Journal* also ran Maj. Gen. John Schofield's report as academy superintendent, in which he dwelt heavily on the Whittaker case. He reiterated that West Point was not the place to enforce social equality because the forced association "destroyed any disposition" white cadets had toward Black cadets—noting that this was no doubt due to the "bad personal character of some of the young colored men sent to West Point." He went further, saying, "To send to West Point for four years' competition a young man who was born in slavery" was akin to having a "common farm horse be entered in a four-mile race against the best blood inherited from a long line of English racers." Simply put, Schofield said, Black cadets were not equipped to compete with white cadets. Regarding segregation in the army, Schofield was quick to point out that he had been one of the first to call for allowing Black troops in the army in the Civil War. However, he differed from Sherman, stating that it was not a question of constitutionality but rather a question of social relations: "The citizen does not forfeit his right to social liberty any more than to religious liberty by becoming a soldier."[51] In this he did not seem to take into account Black cadets' social liberty, only the social liberty of whites to deny social equality to people of color. Unlike Sherman, Schofield made no mention of the Constitution but deferred to social norms based on the supremacy of whites.

In January 1881, Whittaker received the court-martial he had requested, which was covered in detail in the pages of the *Journal.* At the same time, Schofield received an appointment to the Southern Division and Maj. Gen. Oliver Howard took over as the superintendent at West Point. This was seen by the editors of the *Journal* as a political move punishing Schofield for his treatment of the Whittaker case, although Sherman categorically said that it was not.[52] Howard was expected to make major reforms at the academy, which Schofield candidly stated "would totally change the character which has given the Military Academy a worldwide renown and placed its graduates among the foremost servants of the Republic."[53] Such open criticism is surprising given that brother officers were loath to openly take issue with each other. However, this was also the era of the Fitz John Porter and G. K. Warren cases, which opened old army arguments from the War of the Rebellion; some acrimony could be expected. Although the overall result of the court-martial and subsequent actions of the president cleared Whittaker of any wrongdoing, he was later expelled from West Point for academic deficiency, having failed an exam in June 1880, only two months after his assault. This, the *Journal* remarked, meant that "Mr. W. is happily not only out of the penitentiary, but out of the service."[54] It remarked in April that the case left an unsavory impression with the public that "a hole can be found to crawl through, in any adverse decision of a military court." The *Journal* worried that public opinion might show that "military law, as administered by officers, contains more flaws than substance."[55]

While the *Journal* undoubtedly represented a consensus of opinion within the army, it is also fair to ask how much of a role it played in shaping that opinion. A letter from Brevet Lt. Col. John Hamilton that ran in June 1880 serves as a good example of many that the *Journal* received during this time. He noted nearly all the incidents of the Whittaker case that the *Journal* had written about, including small details such as Henry Ward Beecher preaching about West Point. It was clear he was an avid reader. And while he was a "Republican in principle" and had "no prejudice against a negro that is not held by a vast majority of my race," he regretted that society was forcing racial integration onto West Point. In his mind, society should solve that problem first: "If senators, preachers, autocratic editors will but eat, sleep, chum with, take into their families negroes . . . West Point will." Additionally, he said, if civilian university students would stop pulling pranks, then West Point cadets would not be tempted to emulate them. The army was not the place for a social experiment so that people could have "just as debased an Army as you wish to legislate or administer it into."[56] The depiction of the Whittaker trial as presented by the *Journal* had absolutely affected Hamilton's perception the intersection of race, West Point, and the army.

From the outset, the introduction of Black cadets at West Point had captured the attention of the *Journal* and the rest of the army. It was when Black cadets were seen as troublesome or outspoken, however, that they attracted the most interest. The Whittaker trial and court-martial stand as the most notable example, prompting public statements from the commanding general of the army and the superintendent of West Point. From their statements and from the letters and editorials in the *Journal*, we can develop a general picture of the army's reaction to Black cadets at West Point.

First, the army remained a rigid, classist organization resistant to change and defensive of its culture. Of senior officers, only Oliver O. Howard dared to buck the popular trend of defending the academy no matter what, from the period of 1870–80. The *Journal*—although it was not run by West Pointers—fell in line with the officer corps in this matter. In this highly stratified, classist environment, it is nearly impossible to ascertain the reactions of enlisted soldiers. If they objected to the *Journal*'s interpretation of events at the academy, they did not express them in the pages of the *Journal*. We have seen that the editors were not averse to running letters that disagreed with their point of view, and the scarcity of such letters speaks volumes for the whole of the army.

Secondly, the army remained rigidly fixed on operating within the social norms of the era and doing nothing that might precipitate broader change. Although most of these officers had fought to eradicate slavery, they clearly believed that Black men were not yet fit to enter the officer corps. Whether they believed the problem to be a social or political one, nearly all were united in the belief that the white race was superior. The notable exception to this rule were Generals Howard and Sherman. Howard was a noted abolitionist, which made him radically progressive compared to his peers. Sherman, although privately believing that social norms needed to change before Black officers could be accepted in the army, also realized that the Constitution required him to sacrifice his private beliefs for the laws of the nation. Even if he wrote his challenge knowing full well that the civil authorities would never accept it, it is still a profound admission of the primacy of the Constitution at a time when most army officers deferred to social norms above the letter and spirit of the law.

In sum, the majority of objections to Black cadets at West Point centered on how the army should not be used to change the social norms of the day. From letters, interviews, and editorials, nearly all opinions concerning Black cadets at West Point focused on how improper it was for Congress and the president to use the army to conduct what was perceived as a social experiment. Authors stated that Black officers would damage readiness, would degrade the quality of life in the army, and would put white wives in danger. Even those who believed that Black cadets should be given a chance resisted any changes to the academy system that were imposed on it from external institutions and forces.

The army's reaction to Black cadets at West Point during the Reconstruction era shows us how the institution's conservative and reactionary mindset toward change has been a constant over the last 150 years. These same arguments would be dusted off and reused to contest desegregation, women's admittance to the service academies, women serving in combat arms, and LGBTQ servicemembers in general. These social changes within the military bear much in common with integration at West Point during Reconstruction, demonstrating repeatedly that, as Sherman said, "the Army would soon grow accustomed to it."

Notes

1. Alfred Terry to William T. Sherman, June 20, 1870, William T. Sherman Papers, General Correspondence, 1837–1891, Apr. 25–Oct. 20, 1870, Library of Congress, Manuscript/Mixed Material, https://www.loc.gov/item/mss398000035/.

2. Joan Boudreau, "A Civil War Press Pass for William Conant Church," American Printing History Association, https://printinghistory.org/civil-war-press-pass -william-conant-church/.

3. Frank Luther Mott, *A History of American Magazines, 1850–1865,* vol. 2 (Cambridge, MA: Harvard University Press, 1938), 39–40.

4. Richard W. Stewart, ed., *American Military History, Vol. 1: The United States Army and the Forging of a Nation, 1775–1917* (Washington, DC: Center of Military History, 2005), 316.

5. Bradley Peniston, "Editor's Note: The Next 150 Years," *Armed Forces Journal,* August 1, 2013, http://armedforcesjournal.com/editors-note-the-next-150-years-3/.

6. Alexander David Lynch, "Armed and Dangerous: The Ascendance of the National Rifle Association" (Senior Project, Bard College, Annandale-on-Hudson, NY, 2020), https://digitalcommons.bard.edu/senproj_s2020/190.

7. Frank Smyth, "The Early NRA Had Nothing to Do with the KKK," *Progressive Magazine,* May 6, 2020, https://progressive.org/latest/early-nra-had-nothing-to-do -with-kkk-smyth-200506/.

8. Boudreau, "Civil War Press Pass."

9. Issues of the *Army and Navy Journal* (hereafter cited as *ANJ*) have been digitized and can be found online: see the Internet Archive, Google Books, and Hathi Trust. See *ANJ* 7.49 (July 23, 1870): 772, https://archive.org/details/sim_armed-forces-journal _1870-07-23_7_49.

10. "The Colored Cadet at West Point," *ANJ* 7.49: 775.

11. "Devilling Plebes," *ANJ* 8.2 (August 27, 1870): 56, https://babel.hathitrust.org/cgi /pt?id=coo.31924069759904&view=1up&seq=43.

12. *ANJ* 8.2: 61.

13. Albert E. Williams, *Black Warriors: Unique Units and Individuals* (Haverford, PA: Infinity Publishing, 2003), 20–22.

14. *ANJ* 8.4 (September 10, 1870): 61.

15. *ANJ* 8.11 (October 29, 1870): 173.

16. *ANJ* 8.11: 173.

17. *ANJ* 8.22 (January 14, 1871): 349.

18. *ANJ* 8.22: 349.

19. *ANJ* 8.27 (February 18, 1871): 429.

20. Elizabeth D. Leonard, *Benjamin Franklin Butler: A Noisy, Fearless Life* (Chapel Hill: University of North Carolina Press, 2022), 175, 183.

21. *ANJ* 8.27: 429.

22. *ANJ* 8.23 (January 21, 1871): 364.

23. "The Army and West Point," *ANJ* 8.28 (February 25, 1871): 444–45.

24. *ANJ* 10.10 (October 19, 1872): 153, https://www.google.com/books/edition/The _Journal_of_the_Armed_Forces/dWtFAQAAMAAJ?hl=en&gbpv=1&pg=PA1 &printsec=frontcover.

25. "Various Naval Items," *ANJ* 10.19 (December 21, 1872): 294.

26. *ANJ* 11.48 (July 11, 1874): 761, https://babel.hathitrust.org/cgi/pt?id=coo .31924069759938&view=1up&seq=1.

27. Leonard, *Benjamin Franklin Butler*, 191.

28. *ANJ* 12.43 (June 5, 1875): 687, https://archive.org/details/sim_armed-forces-journal _1875-06-05_12_43.

29. *ANJ* 14.46 (June 23, 1877): 739, https://archive.org/details/sim_armed-forces -journal_1877-06-23_14_46.

30. *ANJ* 15.12 (October 27, 1877): 180, https://babel.hathitrust.org/cgi/pt?id=coo .31924069759979&view=1up&seq=7.

31. *ANJ* 16.27 (February 8, 1879): 480, https://babel.hathitrust.org/cgi/pt?id=coo .31924069759987&view=1up&seq=8.

32. "Colored Cadets," *ANJ* 15.27 (February 9, 1878): 427.

33. "A Strange Affair at West Point," *ANJ* 17.36 (April 10, 1880): 729, https://babel .hathitrust.org/cgi/pt?id=coo.31924069759995&view=1up&seq=7.

34. "West Point Cadet Case," *ANJ* 17.37 (April 17, 1880): 747–49.

35. "The Case of Cadet Whittaker," *ANJ* 17.37: 756.

36. "Affairs at Washington," *ANJ* 17.37: 758.

37. "The Whittaker Case," *ANJ* 17.38 (April 24, 1880): 778.

38. "Military Methods," *ANJ* 17.38: 780.

39. "Retiring Non-Commissioned Officers," *ANJ* 17.39 (May 1, 1880): 796.

40. *ANJ* 17.40 (May 8, 1880): 816.

41. See Rory McGovern, "'You Need Not Think You Are on an Equality with Your Classmates': Resistance to Integration at West Point."

42. *ANJ* 17.38: 799–800.

43. *ANJ* 17.44 (June 5, 1880): 904; *ANJ* 18.3 (August 21, 1880): 42, https://babel .hathitrust.org/cgi/pt?id=coo.31924069761603&view=1up&seq=5.

44. "West Point a Typical Institution," *ANJ* 17.50 (July 17, 1880): 1028–29.

45. *ANJ* 18.7 (September 18, 1880): 123.

46. "General Sherman's Report," *ANJ* 18.16 (November 20, 1880): 317.

47. "General Sherman's Report," *ANJ* 18.16: 317.

48. Indeed, during the 1870s many promoters of Black equality became advocates of segregation in the army as it at least guaranteed Black service. Many worried that integrating the military would result in white recruiters simply selecting only white

recruits. Ben Butler noted in 1876 that "as soon as [Black regiments] are done away with, they will refuse to enlist any more colored men and so there is no way for a colored man to compel a recruiting officer to take him, the colored soldier would be a thing of the past." Historian Arlen Fowler states that Sherman's support for former Union Maj. Gen. Ambrose Burnside's failed 1878 bill to integrate the army was based on his desire to eventually rid the army of Black soldiers altogether. See Fowler, *The Black Infantry in the West* (Norman: University of Oklahoma Press, 1996), 119.

49. Leonard, *Benjamin Franklin Butler,* 223.

50. "Gen. Sherman's Ideas: The Proposed Captain-Generalship for Grant," *Washington Post,* December 23, 1880, 1.

51. "Maj.-Gen. Schofield's Report," *ANJ* 18.16: 317.

52. "Popular Gossip about the Army," *ANJ* 18.23 (January 8, 1881): 454.

53. "What General Schofield Says," *ANJ* 18.23: 455.

54. *ANJ* 19.34 (March 25, 1882): 766, https://babel.hathitrust.org/cgi/pt?id=coo .31924069761611&view=1up&seq=642.

55. "Sergeant Mason's Case," *ANJ* 19.35 (April 1, 1882): 788–89.

56. "What West Point Must Do," *ANJ* 17.46 (June 19, 1880): 948–49.

THE INTEGRATION OF USMA IN THE POPULAR PRESS, 1870–1871

Amanda M. Nagel

The front page of the Washington, DC, *Evening Star* on March 8, 1870, included a two-sentence paragraph: "General [Benjamin] Butler yesterday appointed Charles Sumner Wilson, of Salem, Massachusetts, a colored boy, to a cadetship at West Point. This is the first colored boy ever appointed to a cadetship."[1] That exact wording also appeared six days later in the Charlestown, West Virginia, *Virginia Free Press*. By March 17, 1870, Frank H. Fletcher's letter to the editor of Washington, DC's *New Era* (an African American publication) concerning the potential appointment of Wilson to the United States Military Academy (USMA) appeared on the paper's front page. Fletcher, from Salem, Massachusetts, was "surprised and gratified" to hear that General Butler proposed Wilson for a position at the academy. According to Fletcher, Wilson's father Thomas served in Company F, 55th Massachusetts, and was described as "a seaman, not a native of the United States." Fletcher said Wilson's mother Rebecca was an abolitionist, intelligent, and "a native of Ransom, Massachusetts, a townswoman of the lamented and eminent Horace Mann." Fletcher asserted that Wilson, "of quite direct African descent," excelled in school and would be a great candidate for USMA. Most significantly, Fletcher said, "Appointment of colored cadets will become a matter of course in a short time. But his success in any case must be in a faithful application to master and acquire the peculiar education."[2]

However, despite General Butler submitting Wilson for a cadetship due to his high academic performance (he was among the top in his class in Salem) and because he was the son of a Civil War veteran, Wilson was also only seventeen at the time of appointment. Therefore, while Wilson was the first Black man to receive a congressional nomination, he would not be eighteen by the time of his appointment and could not accept the nomination to USMA. Additionally, some even mocked Wilson's conditional appointment, claiming other states would follow suit by appointing a woman or an Indigenous American to USMA, something completely preposterous to white men at the time.[3] By April, the *Evening Star* reported that Tennessee congressman William F. Prosser had appointed Alonzo Napier to USMA, claiming Prosser would outdo Butler with his successful nomination of a Black cadet.[4] That same month, the *New Era* confirmed that Wilson was too young for admittance to USMA and instead placed its hopes in Napier's appointment. The same article bluntly asserted the effect that integration might have on USMA: "It is proper that Jeff. Davis and Robert E. Lee, who devoted the education bestowed upon them by the bounty of the Government to efforts for establishing a slave empire, should be succeeded in the Military Academy by a member of the race which they despised and trampled on. The race which contributed many thousands of soldiers to swell the ranks of the army that defended the Union against the late rebellion is entitled to participate in the benefits of the Military and Naval Academies."[5] While it is unclear from which specific newspaper this article originated, the sentiment reflects the sectional differences prevalent during Reconstruction. It also shows that some northerners started to view USMA as "the nursery of loyalty" after the Civil War, despite the academy commissioning hundreds of officers who betrayed their oaths of office to join the Confederacy from 1861 to 1865.[6]

These articles discussing the potential integration of USMA also indicated interest in what would happen to the US Army after the Civil War, what impact emancipation would have upon society after 1865, and how far civil rights would extend after the enactment of the Thirteenth, Fourteenth, and Fifteenth amendments. More specifically, white newspapers and journalists from 1870 to 1871 provide insight into the sentiments of white American society as the country moved on from the Civil War, displaying the complexities of white public opinion that both cheered on the successes and reveled in any failures of Black cadets during the Grant administration.[7]

This close examination of 1870 and 1871 reflects the sectional tensions and their influence on white public opinion in white mainstream press coverage of integration efforts at the academy. This uneven, sometimes sporadic, and frequently inaccurate reporting by white mainstream newspapers led to limited (or no) information regarding the cadets, or an intense focus on Black cadets with constant errors rarely corrected by reporters or editors. However, white newspapers writing about integration undoubtedly influenced public opinion about Black cadets at USMA and, at least among northern newspapers, indicated that support for Reconstruction was not exhausted in the early 1870s, as it would be by the end of the decade. As a result, three distinct perspectives emerged between 1870 and 1871 illustrating the sectional tensions influencing white public opinion on integration at USMA: the fading memory of the emancipationist legacy of the Civil War; southern whites asserting that segregation should be the norm due to Smith's resistance to racism at USMA; and the northern white press who blamed institutions and a small number of cadets for incidents, absolved USMA instructors of their role in Smith's experiences, and avoided acknowledging the racism permeating the United States during Reconstruction.

Initially, many journalists mused over the identity of whoever might be the first Black cadet and which congressmen would have the distinction of succeeding in their nominations. In part, this is why so many early articles focused on Wilson, then Napier and Michael Howard. James Webster Smith, who would have that very distinction the white press focused on, was not mentioned until June 1870. Every so often, newspapers like the *Memphis Daily Appeal* included single sentences on the topic as part of their broader news coverage, in one instance indicating that Rep. George C. McKee planned to appoint an African American cadet to USMA from his district.[8] However, it was not McKee but Rep. Legrand W. Perce, also of Mississippi, whose appointment of Michael Howard was successful. Howard's father was Merrimon Howard, one of the first Black legislators elected to the Mississippi state senate. Similarly short in their prose, other newspapers like South Carolina's *Charleston Daily News* paired mention of Howard's appointment with speculation that there might be "several" cadets successfully appointed for the next academic year, as multiple congressmen had released information to the press about their desire to appoint African American cadets.[9]

However, some brief missives were not so benign in their discussions concerning potential Black USMA cadets. In mid-May, the *Memphis Daily Appeal* noted, "It is related that, when one of the Radical appointees to West Point was asked how the colored cadets would be received, he replied: 'They would not have lived two nights after they entered.'"[10] In late May the Jackson, Mississippi, *Weekly Clarion* went even further, referring to Perce as a "carpet-bag Congressman." The publication claimed to be "surprised that his [Howard's] Radical friends do not present him with an elegant cadet's uniform. We trust he will start soon for West Point, for we are anxious to see how this embryo colored major-general will be received by the pale-faced cadets."[11] This clear disdain indicated that even though white newspapers in the South would follow the story, their coverage would be minimal unless some sort of controversy or failure occurred that would then confirm their readers' preconceived notions about African Americans.

In late May 1870, newspapers in New York, with a much wider readership, began reporting on Howard's arrival at West Point, resulting in distinct interpretations of what would occur due to integration. The *New York Sun,* a paper that geared toward a working-class audience and was known to dramatize stories, spun a particularly unsympathetic tale of Howard's first day at the academy. Aside from calling the new cadet by the wrong name ("Cadet Master Charles Howard"), the *Sun* claimed that "the absolute arrival of an African, commission in hand, is too much for West Point human nature to endure. Aristocratic professors and jaunty cadets are speechless. The time for the breaking forth of their indignation has not yet arrived. They cannot do the subject justice, but their indignant countenances and ominous looks indicate the coming storm." The paper described Howard's speech as "decidedly of the plantation," his appearance healthy enough to not be rejected by a medical examination, and impertinent both when he was denied a room at the West Point Hotel and when addressing the problem with Col. Henry M. Black, USMA Commandant ("Colonel Jerry Black" in the *Sun*). Reportedly, "a very serious council of war" was called, where Howard requested that his rights not be violated, only to be waved off by the commandant. The *Sun* mentioned that white cadets expressed frustration, with a few threatening to resign, while others reportedly desired to lynch Howard. Ultimately, the *Sun* believed that Howard would "fail in mental examination, and go back to Mississippi," in part because Colonel Black and Gen. Edmund Schriver, USMA's inspector general, "are opposed to the African"; therefore, "the black boy will remain on the plantation." The article also mentioned a second Black cadet to be appointed by Judge Solomon L. Hoge of South Carolina, followed quickly by a dig at General Butler's attempted appointment, because Wilson "was too young, and the General knew it. He only appointed him for political purposes."[12]

In comparison, the *New York Times* addressed Howard's arrival in a more measured and detailed story. Their correspondent detailed that Howard, upon arrival, discovered that he would have to find boarding with a local Black family as reportedly "there were no accommodations left" at Cozens' Hotel. Additionally, Howard's reception among cadets and military faculty was mixed, with many unsure of what might occur if Howard passed the entrance examinations. The correspondent noted: "Even the officers, while discarding all political bias in the matter, and after having fought for the colored race both at the polls and on the battle-field, while feeling kindly toward him, speak doubtfully of the expediency of this venture. They regard it as a more

decided advance of the colored man into social circles than even the eleva-
tion of Senator [Hiram] Revels to Congressional rank." Unlike the *Sun,* the
Times correspondent expressed at least some empathy with Howard regarding
his potential isolation as a cadet but thought that Howard "will be treated,
by both officers and cadets, with courtesy and kindness, their sense of honor
being too high to permit any other course."[13] In a separate snippet from the
"News of the Day," further sympathy for Howard appeared:

> If reports speak truly the enterprising colored lad may expect to have
> something more than the stated curriculum of our military school to
> pass through, and the resolute expression which appears to be visible
> on his countenance will probably find occasion to be deepened should
> he succeed in commencing his studies. Let us hope that the scruples of
> the officers in command will not unduly affect the severity of the pre-
> liminary examination, since it would be somewhat difficult to explain
> why a race that has shown its fitness to occupy the Bar, the Bench, the
> Pulpit and the seat of the legislator, should be debarred from military
> command.[14]

Such disparate reporting between two New York newspapers indicates varied emotions among white Americans regarding the integration of USMA.
Moreover, what each publication chose to include (or ignore) in their coverage provides insight into how they sought to sway their readers' opinions. Despite the *Sun* shifting away from its previous focus on tabloid-like, sometimes
falsified, news narratives toward society and human-interest stories, tabloid
sensationalism still influenced its journalists' writings. In contrast, as the *New
York Times* had quite a wide circulation well before 1870, it is not surprising that its journalists would emphasize facts and err on the side of empathy,
however briefly. The newspaper had a reputation for measured and legitimate
news coverage, and these articles only reinforced *Times* journalists' attempts
to maintain that reputation while also shaping public opinion in favor of integration at USMA.

When local newspapers across the nation mentioned very little to nothing about Howard's arrival, one can surmise that white popular opinion in the
area was either indifferent to integration or opposed it; therefore, limited
coverage existed. The latter was the case for South Carolina's *The Charleston
Daily News.* On May 27, 1870, that paper printed two sentences regarding
Howard's arrival: "A West Point special to the Sun says the negro cadet from
Mississippi has arrived, creating great commotion. The hotel refuses to en-
tertain him."[15] The *Nashville Union and American* shared similar sentiments,

describing succinctly how the examinations would begin in early June and claiming that if Howard "proves to be three-fourths black he is certain to be rejected under the rules governing the medical examination now in force."[16] However, the *Alexandria Gazette and Virginia Advertiser* seemed more apt to critique USMA itself than Howard. Their journalist noted that cadets chosen "are hereafter to be exclusively selected from the sons of army officers," regardless of race. Their only mention of Howard was that he arrived at USMA for examinations and that "his arrival created quite a 'sensation.'"[17] For the Augusta, Maine, *Daily Kennebec Journal,* Howard "was received rather coldly by the cadets," but he "persevered, and gave promise of being able to fight it out on his own line."[18] However, by such scant reports, the paper's readers must not have been interested in the story; therefore, it was given short shrift.

Coverage, however lengthy or succinct, continued into the examination season, and finally some newspapers drew public attention to the second African American cadet appointed to USMA: James Webster Smith. A *New York Herald* correspondent claimed to interview Howard and Smith about their first few days at USMA. As a disclaimer of sorts, the journalist pointedly mentioned, "West Point is a place where newspaper correspondents are only tolerated," a widely shared sentiment among many journalists during Reconstruction. The correspondent added that Howard "most emphatically" rejected reports of his mistreatment so far at the academy. Instead, Howard said his only complaint was isolation. This was all the correspondent mentioned regarding the interview. The article instead focused on how white cadets responded to integration and described Howard's intellect, character, and capabilities, and denigrated these in a variety of ways. The correspondent even said:

> The indications are that if they declare the colored boys disqualified, according to all the tests usually made with white plebes, they will be denounced by certain powerful political cliques as having succumbed to a "copperhead" sentiment, and if the boys are pronounced acceptable other almost equally powerful parties will say that they did not have the courage to face the radical music. It is easy to be perceived, therefore, what a strait the faculty is in; but from what I know of them I can safely assert that they will do their duty in the matter fearlessly, without fear or favor.[19]

However, the *New York Times,* with its wider circulation, called out the *Herald,* as well as other newspapers, for false reporting in its interview with Howard and Smith, implying that public opinion should not be swayed by that article. Instead, the *Times* correspondent sought to balance the coverage and encourage patience and understanding as USMA integrated. The journalist said that the environment "has been more of perplexity and anxiety than anything else in the effort to divine the policy that should prevail, and to mark the line between the social and the official treatment." To this reporter, the most challenging event to determine that line was dining. After Colonel Black ordered equal treatment of all cadets, some of the white waiters had to be "informed that unless they were willing to serve all the guests of the table alike they must withdraw. They have remained." The correspondent noted that in all other ways, from salutes to duties to barracks locations, Howard and Smith were reportedly treated equally to white cadets. The two were supposedly "greatly favored," "treated with the most generous courtesy," and their peers "resolved that any act of discourtesy or unkindness manifested by any of their number toward the colored comrades would be condemned, and the offender discountenanced."[20]

Even so, white perceptions of race during the late nineteenth century influenced how the *Times* correspondent discussed Howard and Smith. Like other papers, the *Times* called Howard by the wrong name and claimed he had only "fifteen months' education," and therefore was not qualified for a cadetship. The journalist also claimed Smith "seems robust, and is well shaped, but there is a peculiarity in his eyes. Altogether, he does not appear hardy, and shows suspicious taints of Mongolian blood." Additionally, the correspondent claimed that the two men would be on probation first, and then most likely not pass examinations because they were not on par with their fellow white cadets. For the reporter, "It would seem, from present evidences, that if they are judged by the regular standard, they cannot be admitted. Then, unless political pressure is used, they must be retired for a while at least."[21] No matter the patience and understanding encouraged at the start of the article, the latter half starkly showed how far that patience and understanding went. It was akin to white support for abolitionism, in that whites would encourage ending slavery but not always champion citizenship and civil rights for freed people in the spirit of true social equality.

Such a distinction in language is most glaring when compared to Black newspapers like the *New Era*. It is not surprising that African American communities and newspapers would be candid regarding integration at USMA, and particularly its impact on both Howard and Smith as well as white cadets:

Up to this time the "white man" has had that institution all to themselves—or, rather, a few privileged whites—and the cadets now there seem to have inherited from old slave-driving times the same idea. At any rate, they can't comprehend that a colored boy has any rights they

THE INTEGRATION OF USMA IN THE POPULAR PRESS

are bound to respect.... The law makes no distinction between colored and white boys—opening the institution alike to both. Will they be permitted to drive this young man from West Point for the only reason that he is guilty of having "a skin not colored like their own."[22]

Outside the possibility of editorials, no white journalists would have taken such a stance on integration and how the reactions to it were interwoven with racism prevalent in the United States. As a result, the mainstream newspaper coverage of USMA's integration reflected the wider white American public's attitudes regarding Black cadets and Black officers in the US Army.

The white press noted reports that both Howard and Smith failed their examinations, the former for failing the written exam and the latter the physical one. According to the *Chicago Tribune,* "The Inspector General declares that the color of the skin had nothing whatever to do with either rejection. He says the young men messed with other candidates, were under the same rules and protection, and have been uniformly treated with kindness by the officers and cadets alike."[23] Philadelphia's *Evening Telegraph* provided more information by reprinting an article from the *New York Tribune.* The article noted that some newspapers critiqued the USMA Board of Examiners for failing Howard and Smith, then provided the alleged justifications published by General Schriver's office. Howard's rejection was due to his limited education before entering USMA, in that "he had not been at school over a year" prior to his arrival. Therefore, he failed the written examination, but easily passed the physical test. Smith's supposed rejection was due to his eyesight, "an affection of the lungs, and it is well known that he is generally in bad health." However, since Smith had attended Howard University prior to USMA, the journalist believed Smith would easily pass the written test, "but the law of Congress prevents his admission on account of the physical disability." Additionally, the article claimed that the typical cadet hazing of plebes did not happen to Howard and Smith "because the cadets thought the people would say they were roughly handled because they were colored boys." Finally, the correspondent asserted that because most white cadets and officers at USMA were Republicans, it meant that no matter what was published in newspapers, mistreatment of Howard and Smith could not happen due to the "preponderance of Republicans" there.[24]

However, at least two newspapers ridiculed Smith and Howard, asserting their alleged rejections were warranted due to their race. While also reporting the supposed rejections, the Columbia, South Carolina, *Daily Phoenix* noted, "These boys were rejected because they were black, and the colored people will so understand it. You can't put your finger in the eyes of black men and tell them it is raining, even if they were once slaves." Along a similar tone, the Ashland, Ohio, *States and Union* said, "Choking to death with melted butter is not the only way to kill off inconvenient people."[25]

Additionally, these newspapers published incorrect information and did not check the regulations governing physical requirements for military service nor the timeline of examinations before printing and distributing the information. Such actions were indicative of a desire to see both potential cadets fail and for USMA to remain as it was, with no cadets from minoritized peoples. Instead, as reported by the *New York Herald*, Howard passed his medical examination and had to still take the academic examination. While the correspondent said that most likely Howard would not pass due to limited schooling prior to USMA, he still had the opportunity to take the examination. In Smith's case, while he had an eyesight issue, it was not enough to reject him on medical grounds. He was first examined on June 1 and placed on probation. His second examination on June 5 resulted in lifting that probation.[26] Smith would be allowed to sit the academic examination, and if he passed, he would become a cadet. However, inaccurate reports in newspapers asserted that Smith was still on probation and that "he remains at the Academy for a whole year, during which time every effort is made to cure whatever physical defect he may have. If he does not improve by the end of the twelve months he is then rejected."[27] Neither cadet had been rejected as of mid-June, yet newspapers reported such information as fact.

The *Herald* lambasted the *New York Times* in return for publishing false information and an editorial that it considered too lenient on some at the academy, while also highlighting how significant these possible cadetships were to USMA and the nation. Like other papers, the *Times* reported on Howard and Smith's alleged rejections. The *Times* editorial mentioned by the *Herald* asserted that African Americans "must put their best men to the front" when seeking racial uplift, since the entire race would be judged by successes or failures of a few. The editorial also noted, "No blame for this, of course, rests on the authorities of the Academy; and none should rest on the colored race."[28] In response, the *Herald* asserted, "No one will deny that this colored cadet question is one of great moment." The journalist also remarked that the faculty were aware of the scrutiny upon USMA and supposedly sought to minimize any potential conflicts among the cadets. The *Herald* also mentioned that Howard and Smith reportedly faced less hazing than other new cadets and that all the white cadets "have to a man acted as men should in the matter." Therefore, exaggerations or falsehoods did "gross injustice to the officers in the Academy," who were treating Howard and Smith like white cadets: "Justly, honorably, without fear or favor."[29]

To that end, the *New York Sun* encouraged readers to focus on how USMA's process would ensure impartiality despite how white cadets responded to their peers. Its correspondent spoke with Col. Edward C. Boynton, USMA's quartermaster, who confirmed that the exception for Smith regarding his eyesight was typical for many applicants and that impartial grading by assigning cadets a number rather than having their name on the examinations ensured each cadet had the same chance to succeed. While the correspondent noted the faculty's potential for impartiality, they also documented how white cadets responded to Howard and Smith. The journalist reported hearing white cadets using racial epithets to encourage Howard and Smith to leave USMA, asserting that if either chose to respond, "they would have to fight two-thirds of the Academy." The correspondent referred to their stoicism in the face of such treatment as "Christian resignation." The correspondent also reportedly had an interview with the two, where they provided family backgrounds as well as notable connections like David Clark, philanthropist and former US Army officer of Hartford, Connecticut; and Mississippi senator Adelbert Ames.[30]

However, the *Sun* also asked New York senator Roscoe Conkling, Adm. David Dixon Porter, and Secretary of War Gen. William W. Belknap about integration at the academy; their reactions were varied. Conkling asserted, "They are not representative boys at all. They do not represent any race." Therefore, Conkling believed their candidacies were "a farce," that both would fail examinations, and that "the whole turmoil will have been for nothing, except that these boys will have been slaughtered forerunners of a great reformation—a reformation which gives representation to four millions of new citizens." Admiral Porter claimed he liked African Americans enough, "in their places," but believed USMA was "no place for them." Porter also said that he thought white sailors would "drop them overboard on the first cruise." It was Belknap whose opinion was the outlier here: he thought they would be treated like any other cadet, despite existing prejudices in the United States. Belknap said, "There is a tremendous public sentiment against the race. West Point is naturally aristocratic, but it must come under. Four millions of enfranchised men are entitled to one-tenth of the offices." However, Belknap agreed with Conkling that Howard and Smith did not seem to be "representative" but expressed hope that they would succeed, and he would be proven wrong.[31]

That hope for success also appeared in the *Chicago Tribune* in its news from Washington, DC, section, and included critiques of people speaking on the two cadets' behalf. The short snippet discussed Secretary of War Belknap's visit to USMA from June 7–16, 1870, where he spoke with Smith and Howard. Belknap asked "if they had fair play," and both said yes. After prodding, they told Belknap about the racial epithets overheard from other cadets speaking amongst themselves but never directly aimed toward them. Here, the *Tribune* shifts and provides its readership a particular perspective: "This [the racial epithets] was the whole arm of the basis of the tale afloat that the boys were kicked and cuffed. It was a tale set afloat and believed by that large class of disagreeable people who did not want the war to stop; who want to revive it again, and keep it going." Nevertheless, the article concluded with Belknap's speech to the graduating cadets, where he asserted that very few of USMA's graduates had ever "borne the badge of dishonor," with most graduates remaining loyal to the United States when secession occurred.[32] Thus, while seemingly supporting the Black cadets, the *Tribune* also indicated weariness for discussions of equality and rights only five years after the Civil War ended.

Others used integration at USMA to condemn both West Point graduates who had joined the Confederacy and racism in general. This included the Clearfield, Pennsylvania, *Raftsman's Journal.* In a quick note, it mentioned how "Democratic contemporaries" expressed displeasure with Howard and Smith attending USMA. However, the publication then said, "They [former Confederates] have not yet forgiven the colored people for helping to crush the rebellion, and cannot endure the thought of permitting colored officers to fight for us in any future war."[33] Likewise, New Jersey's *Paterson Daily Press* referred to "the miserable snobs of West Point," whose interactions with Howard and Smith were "proving how small an amount of manhood and soul can be enclosed in a white skin and pass for a man."[34] A *New York Dispatch* article indicated continued sectional animosity. Referring to USMA as "a nest of aristocracy," the *Dispatch* blamed this "class-pride and assumption" for generating "a rank crop of traitors, North and South, who broke their 'Union oath,' and became ingrates to the country which had nourished them at her breast." The article also addressed interactions between white cadets and Smith in particular, including some particularly sharp language regarding possible prejudice among white cadets for their comrade in arms, calling it "sheer nonsense." Additionally, the *Dispatch* noted that during the Civil War, "in escaping from rebel activity" as fellow prisoners of war and fighting alongside Black regiments, US soldiers "never dared to object to such treatment." As such, "if a negro is good enough to fight beside, he is certainly good enough to drill beside."[35]

The Connecticut *Waterbury Daily American* asserted solid support for Smith as well. They wrote, "West Point to-day is as replete with prejudices and aristocratic tendencies as ever an old garret was with cobwebs and dust." As a result, "the old feelings of caste are a living thing of the present with the cadets" at USMA. Listing Smith's academic accomplishments and noting how white students in Connecticut had treated him as an equal, the *Daily American* thought all that was needed at USMA was for Smith to "have a fair chance" to achieve greatness. The correspondent also argued that Smith was a "representative youth of his race," who stood "in the very advance of the vanguard of his race in this new destiny to which the issues of a great war have raised that race."[36] It is clear that some newspapers encouraged the public to not think so highly of USMA and the officers it graduated, particularly if they might become a traitor to the country or express racism to a brother-in-arms while training.

Beginning in July 1870, articles, typically short in length, regarding the results of the examinations were published, mostly focused on factual information, while others provided commentary on the results of USMA's examination period. Here, too, reporting is muddled. Many papers claimed that three Black cadets were sitting examinations and that all three were rejected. Some continued to insist that Smith was on probation (or "sent to hospital") for one year before becoming a full-fledged cadet.[37] Other outlets like the Burlington, Vermont, *Weekly Free Press* noted that Smith was indeed accepted to the academy, dryly asserting in reference to his admission, "We presume the school will go on as before."[38]

But public attention quickly turned toward a letter written by Smith to David Clark and published by Clark, regarding Smith's experiences at USMA. Multiple newspapers reprinted the entire letter, with a few providing commentary alongside it. The letter garnered so much attention because it articulated so clearly what Smith, in his own estimation, experienced in his first month at USMA. Smith mentioned contending with examinations and "insults and ill-treatment" from white cadets, along with further isolation after Howard departed. He also detailed some of the cadets' attempts to encourage his departure from USMA: making sure he did not sleep; trying to deprive him of food at mealtimes; refusing to stand next to him at drill; and finding ways to ensure he might be punished for not knowing some movement during inspection. Smith even considered quitting because of this resistance and hostility, despite not wanting to leave USMA. He asserted: "If I complain of their conduct to the commandant I must prove the charges or nothing can be done; and where am I to find one from so many to testify in my behalf?"[39] Other publications mentioned the letter and its contents but did not republish it.[40] One publication, however, commented that white cadets who were "dissatisfied with the regulations the government has seen fit to establish" should resign, because their behavior toward Smith was "ungentlemanly."[41] Another newspaper pointedly asked why President Grant, General Butler, or any other politicians had not stopped this conduct already, asserting: "The shame and disgrace that attends the white cadets in their babarous treatment of young

Smith will attach to those who have authority at the academy, very soon, if this sort of thing is not stopped."[42]

Interest in the letter led directly to requests for an investigation into USMA and its inner workings, driven by members of Congress. Some of that encouragement might have come from various newspapers publishing Gen. Oliver O. Howard's letter to the editor accompanied by a copy of Howard's letter to Smith, written to the cadet after seeing news of his troubles at USMA. General Howard did so because he believed it might "influence high-minded cadets in his [Smith's] favor," asserting that those "who will persecute a man because they can do it with impunity, will hide their heads when the indignation of true men is excited against them." Howard also had a few choice words for his alma mater. If USMA could not properly protect a cadet, they would struggle to stop critiques coming their way. His letter to Smith encouraged the young cadet to "never think of giving up while you have health to stand the storm" and to withstand the treatment "without any show of fear."[43] The *New York Tribune* said that it hoped General Howard's words "may have the effect on his [Smith's] white comrades that the General desires. It is mortifying to know that there exists such a spirit of intolerance on their part as is revealed in the letter of the persecuted cadet."[44] Other papers, like the Canton, Ohio, *Stark County Democrat* thought Howard was wrong for speaking "loftily about the dishonorable conduct" of white cadets. In its estimation, "If the authorities want to educate negroes at the Academy, proper arrangements should be made for them. Give them a separate table and class rooms and drilling grounds by themselves, and the white boys will not interfere with them."[45] Additionally, South Carolina's *Yorkville Enquirer* expressed its dismay that Smith's cadetship continued past examinations and that white cadets now could only "prove their superiority" through efforts in the classroom. The publication also believed that white cadets had no "right to complain of negro competition, as it has been achieved solely by the efforts of their political friends."[46]

Due to the press's interest in Smith's cadetship, Massachusetts representative Benjamin Butler proposed a House resolution to create a committee to investigate what occurred at USMA. Although a fellow representative motioned to adjourn, which was then carried by the House, Butler said he would propose the resolution again.[47] Massachusetts senator Charles Sumner also "offered a resolution instructing the Military Committee to inquire" about Smith's treatment at USMA and included "the power to sit during the recess" to the committee so it could accomplish that investigation.[48] It was Butler's proposal that eventually succeeded and created a committee to investigate USMA's cultural environment.[49]

When coverage of this inquiry appeared in the press, white public opinion overwhelmingly turned against Smith and his cadetship. As the War Department began providing information, correspondents stated that Smith claimed the letter published was not the one he wrote but a "garbled copy." Also, the drillmaster referenced in said letter denied the accusations, as did six witnesses. The white mainstream press therefore concluded that the letter's contents were an exaggeration of actual events. Belknap ordered a court of inquiry to investigate further, consisting of Lt. Col. James H. Wilson and Majors Henry L. Abbot, Theophilus F. Rodenbough, and Thomas F. Barr.[50] This was after other newspapers reported that before the inquiry, Smith attempted to resign his cadetship, which Lt. Col. Emory Upton refused; that a cadet refused to sit next to Smith in the mess hall, resulting in his arrest; and another cadet refused to drill with Smith.[51] West Virginia's *Wheeling Daily Register* did not mention the investigation but republished Smith's letter with some commentary and paired it with another reprint from the *Philadelphia Sunday Mercury* that supported segregation and rejected Radical Republican politics. The *Register* asserted that white cadets' actions proved "that the Radical idea of equality don't take very well" among the younger generation and that it was "impossible . . . for the two races to mix together with any sort of harmony. It is an evil and an outrage inflicted upon the country by the Radicals for base party purposes."[52]

Claims in white newspapers that the allegations in Smith's letter were exaggerated resulted in more newspapers overtly lambasting Smith and integration at USMA. This included Virginia's *Richmond Daily Dispatch,* whose correspondent compared Smith's cadetship to introducing a new chicken to a brood. The correspondent's conclusion was that due to the reaction of the white cadets (the chicken brood), Smith's "best course is to get out of West Point." The correspondent then immediately added: "But he won't do this. He holds on, encouraged by outside hypocrites and enthusiasts. He writes whining and complaining letters, and makes himself a martyr." Furthermore, the correspondent argued that white northerners were "not ready for social equality" with African Americans, encouraging a separate USMA to educate Black officers, particularly since those white men "fit to be educated to lead the armies of the United States are never going to submit to social equality with the African race, or Chinese, or any of the inferior types of man."[53] Moreover, the *Daily Missouri Republican* asserted that if Smith could not handle the treatment delved out by white cadets, then he should resign.[54] The Jackson, Mississippi, *Weekly Clarion* made clear its perception of civil rights, emancipation, and integration in late July. The paper misleadingly described Smith as a carpetbagger who had moved from Connecticut to South Carolina. It also claimed his treatment was at the hands of white northerners, since "no one can gain admission into the portals of the Academy without making oath that he neither participated in the war against the Federal Government, nor gave aid or sympathy to persons who were thus engaged." Therefore, "the sensibilities of the white race, uncontaminated by fanatical teaching, or lust of office, instinctively revolt against the amalgamation doctrines of the party in power." The article concluded by wryly asking if Republicans really had African Americans' best interests at heart when they supported placing Smith in such an antagonistic position.[55] The *Clarion* chose to omit the fact that Smith grew up in South Carolina, his father was an alderman in Columbia, and that by 1870 former Confederate states were also sending cadets to USMA once again, therefore blaming white northerners entirely for Smith's treatment was a stretch of the facts, at best.

Quickly, white public opinion became focused on results of the investigation, with many publications predicting Smith's court-martial conviction in late July. The *New York Times* thought that due to such a wide "variance with the statements" of those involved, Smith "will be convicted of gross exaggeration or worse."[56] The *Richmond Daily Dispatch* had much stronger language regarding Smith and his enrollment at USMA when reprinting an article by correspondent Donn Piatt, a Civil War veteran who worked for the *Cincinnati Commercial.* Piatt noted that as civil rights "reached the little school on the Hudson," integration would have the same effect as it did on treatment of Black senators: no one was "bound to associate with him." Additionally, while painting USMA as an extension of a supposed American aristocracy, with wealth and privilege on display, Piatt argued that introducing someone with Smith's background—the "son of a slave, the representative of a servile race"—to USMA meant "you can appreciate the convulsion that has come to West Point." In Piatt's estimation, Smith "slipped through, and he remains there like a flaw in a diamond," but Piatt believed Smith should remain at USMA because it "will be so tarnished that we can hope for a reform."[57] For papers like Illinois's *Cairo Daily Bulletin,* such courts-martial and placing some cadets under arrest meant that "Radicals are determined to press matters and compel a recognition of the social equality of the negro race."[58] In Mississippi, the *Weekly Clarion* asserted that this was "the beginning of a series of other troubles which will follow in its train." It turned to a *St. Louis Republican* article to support this assertion. In so many words, that article, reprinted in the *Clarion,* denied African Americans' citizenship, spoke of the fear that might come from having Black officers lead white soldiers and move their way "up this ladder of promotion," and warned that this would then lead to Black men in positions of power if ordered to occupy a location under martial law.[59]

The first court-martial Smith faced was not until October 1870, months after the court of inquiry finished its investigation and another incident occurred in late July. Before this, USMA handled the punishments for insubordination (refusing to drill with or sit with Smith). The investigation revealed that while Smith's statements reflected "some ill-treatment," the contents of his published letter "are not in the main correct." The court also determined that faculty members treated Smith equally to white cadets and that he and another cadet should be reprimanded rather than court-martialed.[60] According to some newspapers, the exaggerations in Smith's letter had to do with influential people around the cadet "who worked upon his mind to transform himself into the position of a representative martyr and pioneer of his race."[61] By late August another investigation began, this time involving Smith and Cadet John W. Wilson, who got into an altercation over the water tank. Smith wanted to fill a bucket for those on guard duty, but Wilson would not let Smith ahead of him in line. Smith reportedly attempted to stand so close to Wilson as to force him away, causing Wilson to strike Smith, and Smith to respond in kind. Both cadets were placed under arrest and an investigation began. The *New York Sun* sought to speak to Smith and Lieutenant Colonel Upton about the incident, but Upton refused, saying, "If you publish anything it will be almost sure to make it worse for Smith among his comrades; it would be better to say nothing about it." Consequently, the *Sun* then sought information from other cadets, which the *Sun* deemed "essentially correct." That same article described Smith as "naturally sensitive, jealous, and resentful," someone who "takes it exceedingly hard," and "sullen, inclined to be disobedient, and quick to quarrel."[62] As a result of this incident, a *New York Times* reprint in the Washington, DC, *Evening Star* declared that Smith "must take care not to lose the public sympathy which has been lavished upon him. . . . Cadet Smith should recollect that his race is, to some extent, on trial in his person."[63]

By mid-October, newspapers began reporting that a court-martial had been ordered, charging multiple cadets in the incident, conveying at least some empathy for Cadet Wilson and criticism for Cadet Smith. A *New York Times* correspondent said, "Wilson has the sympathy of the entire corps. He is a much smaller person than his antagonist, and has always enjoyed the reputation of being a good, quiet, well-behaved fellow. Smith is the exact opposite, his obstinate manner, and many inconsistences having brought him into general disfavor." The correspondent also noted that the corps of cadets asserted that "their antagonism to Smith is not now because of his color, but on account of what they deem his dishonorable conduct."[64] The *Chicago Tribune* also critiqued Smith, asserting that the young cadet was "a failure" and that regardless of "the injurious and unwise influence of politicians, so-called friends of his race, and itinerant 'journalists,' paid by the column for sensational and calumnious articles, it is by many deemed doubtful if he could have made a career at West Point." The *Tribune* joined the *Times* in depicting Smith as stubborn and disagreeable, while simultaneously weak and incapable of racial uplift through what they considered his less than stellar performance at the academy so far.[65] In comparison, other publications mentioned only the court-martial itself and, in some cases, the full composition of the court's members: General Howard, Lt. Col. Thomas C. Dunning, Lt. Col. J. H. Dexter, Maj. Thomas J. Haines, Maj. Louis H. Pelouse, Capt. A. C. Bainbridge, Capt. Michael V. Sheridan, and Maj. William Winthrop.[66]

Even as the court-martial completed its work, and its results delayed until the War Department approved the convictions and sentences, white public sentiment remained interested in Smith's exploits at USMA, but that interest waned by late October. While the *New York Times* summarized the court-martial proceedings almost daily, their commentary was minimal. Their first article on the trial on October 22 listed the charges against Smith and re-capped the testimony of white cadets regarding the first of the charges. The testimony included that of Wilson himself, and his fellow white cadets' stories all corroborated his version of events.[67] The next few articles did the same for the rest of the testimonies regarding both charges, but sometimes the *Times* included commentary on Smith's demeanor, reporting that be became "more calm and self-possessed" as the court-martial process continued.[68]

The final article regarding the court-martial from the *Times,* though, provided the contents of Smith's letter to the court in his defense. Smith's version of events related to the first charge varied from the previous testimony in that he specified that he thought his task of retrieving water fell within the regulations that cadets on guard duty could not be away "from the guard tent for more than ten minutes." That was not the case, but Smith said as a new cadet, he did not know that and sought to return as quickly as possible; therefore, when he ran into cadets at the water tank, he pushed the bucket toward the spout and did not realize it hit Wilson. Wilson then accidentally tipped over the bucket, and Smith thought this was intentional. Smith also said he was almost certain Wilson swung a blow first. On the second charge, Smith adamantly denied the charge of conduct unbecoming. Within this section, the *Times* correspondent included some snide comments, including mentioning how Smith "pathetically alludes to the great burdens that this reproach caused him to bear, the Commandant's kindness and firmness, and the accused's willingness to have any and all testimony developed."[69] The *Times* was not alone with spiteful remarks; Delaware's *Middletown Transcript* referred to Smith as "a regular brute and a discredit to his race" when publishing a short missive on the courts-martial at USMA.[70]

After the War Department reviewed the results of the court-martial, many papers reported that Smith had been acquitted and released from confinement.[71] The initial punishment was to walk the grounds of USMA for six consecutive Saturdays, but the judge advocate of the War Department thought the punishment too light, so it was disapproved.[72] The *New York Times* mentioned the resultant tension among the white cadets. In their estimation, "Smith has not been justified, and they, as a body, have been wronged. The sentiment against Smith is more unanimously bitter and intense than ever."[73] Appearing the same day in the editorial section, an unidentified person asserted that Smith "may yet distinguish himself, if he will control his temper and give close attention to his duties."[74] Tennessee's *Fayetteville Observer* had some harsh words regarding the acquittal. Referring to Smith as "one of the colored pets of the government," the *Observer* asserted that most likely Wilson would face a harsher punishment than Smith, and Smith should "be nominated for the Vice-Presidency, as a reward for his valor in vanquishing so summarily a boy half the size of himself!"[75] However, Connecticut's *Litchfield Enquirer* described the incident resulting in a court-martial thus: "The irrepressible African knocked one of his persecutors on the head with a dipper, to the manifest injury of his scalp and of his feelings as a white person."[76] Indeed, as integration at USMA continued, public opinion remained varied across a spectrum from full-fledged support to vitriol for Smith regardless of circumstances.

Continued sectional strife fed the varied public opinions, embodied by the tensions between the two main political parties. The *New Orleans Republican* noted that Democratic-leaning newspapers initially "delighted" in USMA's integration because Smith's attendance would "degrade the academy." However, Smith's admittance to the academy "enraged the Democrats, who saw peace where they looked for war." Eventually, according to the *Republican,* it meant these papers would claim that after various incidents and Republican support for Smith, somehow it was Democrats who "are the best friends of the colored people."[77] Conversely, South Carolina's *Charleston Daily News* reprinted a *New York World* article that was subtler in its critique of those supporting Smith. It asserted that white cadets viewed Smith as a "favorite," and was treated differently than other cadets because "he has Congressional friends, who are more or less pledged to regard every action unfavorable to him as an act of persecution." The article mentioned President Grant's son, Frederick, also a cadet, who wrote about Smith and USMA to his father.[78] Interestingly enough, though, very little of the correspondence mentioned in the article or information about USMA's integration in the 1870s found its way into *The Papers of Ulysses S. Grant,* eventually bound and published in 2003.[79]

By the end of 1870, the bad press for Smith continued to increase in the white mainstream press, and in December newspapers reported that he was under arrest again and scheduled for a new court-martial. Smith was charged with conduct unbecoming, this time "for submitting an explanation containing disrespectful reflections on the conduct of the reporting officer" after Smith was reported for "delinquency on drill."[80] Multiple articles questioned Smith's integrity, thereby influencing public opinion surrounding the cadet. The *New York Herald* noted that while Smith initially seemed to possess "ability above that of the ordinary negro," his exposure to the USMA curriculum determined he was instead "below the medium ability," and it was questionable if he could finish the coursework and commission. The *Herald,* along with the *Alexandria Gazette and Virginia Advertiser,* claimed Smith possessed a "chronic weakness" of lying, a trait increasingly on display while he was at USMA. As a result, fellow white cadets "quietly ignored him and refused to affiliate with him."[81] Similarly, the Carson City, Nevada, *Daily State Register* asserted, "The early moral education of that youthful Etheopian was probably neglected, and West Point is a poor place to lay the foundation of a gentleman."[82]

Upon news of the new court-martial convening early in 1871, few news correspondents maintained any sympathy for Smith. The *New York Herald* noted the possibility of his expulsion from USMA if convicted, claiming Smith was not expelled after the previous trial due to "the earnest solicitation of members of Congress, who, without distinction of party, asked that Smith have an opportunity of redeeming himself, and that if he misbehaved again they would not interpose in his behalf."[83] Other articles mentioned that Smith's "friends declare it is persecution" but omitted any other information.[84] Iowa's *Buchanan County Bulletin* asserted one of two possibilities existed to explain the second court-martial: "either a systematic course of persecution" of Smith "or he is something of an incorrigible."[85]

However, the *New York Times* still had some kind words for Cadet Smith. After explaining the context surrounding the latest court-martial, the correspondent asserted: "Other parties who are aware of the facts—one of them an eyewitness—states that the whole affair is a conspiracy. With one or two honorable exceptions, the entire cadet corps cherish the most bitter hatred against the color of Smith, and have been eagerly waylaying for the first little possibility in his conduct to cook up a charge" that would result in his expulsion. The *Times* also noted the mental and physical toll that isolation might take on Smith: "Among those 200 sneering companions he is more lonely than a solitary one in the desert. In the gymnasium the others monopolize everything, and, although it may seem incredible, but is a positive fact, that more unpleasant things are said and done to him now than ever before."[86] The following day the *Times* asserted that Smith understood that with the entire corps of cadets opposed to him, as well as those who supported those cadets, "his case is as almost a forlorn hope."[87]

As newspapers reported on the second court-martial, only a few publications offered extensive coverage and in some cases covered Smith's trial alongside other cadets' courts-martial to assert a discipline problem at USMA. Most of the detailed coverage appeared in the *New York Times,* providing readers with a comprehensive summation of each day's testimony. The correspondent also added commentary about Smith's demeanor and professionalism during the proceedings, as well as how the evidence provided by witnesses confirmed that Smith had "been dreadfully aggravated and shamefully abused."[88] Other articles on the proceedings were typically succinct but frequently coincided with reports about separate incidents at USMA—namely, three cadets who had either gone absent without leave or were drummed out by fellow cadets, threatened with tarring and feathering if they returned to the academy. Their experience became central to white public attention, with Smith's experience sometimes included as further evidence of USMA's declining prestige. A few short articles focused entirely on Smith, yet one commented that many other cadets had also been written up for "delinquency at drill," even though Smith had been charged with conduct unbecoming and presenting false statements as well.[89] Interestingly, a *New York Tribune* article used Smith as a short gateway to critique USMA's "lack of discipline in the administration of the post, and a dangerous spirit among the pupils."[90] The *New York Herald* complained of the court-martial proceedings that "it is very doubtful whether even in the old days of slavery this 'bright boy' would have brought as much as the government has already expended on him merely to find out whether he told a lie."[91] As had been the case so far, reporting indicates that white public opinion, influenced by articles like these, remained mixed, but the *Times*' sympathy for Smith became far less common as time passed. After the court-martial ended, newspapers rarely mentioned Smith, focusing instead on whether USMA should continue to exist due to frequent disciplinary issues and on members of Congress who requested the Committee on Military Affairs once again investigate Smith's treatment at the academy.[92]

Ultimately, it was an article in the *Nation* that succinctly encapsulated the complications surrounding integration at USMA during the Grant administration. While the academy was "a nursery of pro-slavery sentiment" prior to the Civil War, after 1865 "anti-slavery agitation" influenced USMA by providing "the ideas of justice, equality, and human brotherhood a lodgment in the national conscience" faster than it would have occurred otherwise. However, integration proved "to be a test which was to reveal the rotten moral condition of the institution." The *Nation* asserted that white public opinion in 1871 left much to be desired in terms of race relations. Those who might treat any minoritized peoples equally were so few that they could "be contained in one room of moderate size," so cadets would have had to express more "self-control" and "moral heroism" than their elders. Additionally, because USMA's structure reinforced constant contact among cadets and isolated them from the outside world, when a cadet was also isolated by his peers, it would create more problems and greater challenges for their success than elsewhere. Notably, "When we see colored people generally rated in society as they ought to be, and as white people are, by their looks, and manners, and education, and character, we shall have a right to be uncompromising in our condemnation of the position of the white cadets towards Smith, but hardly any sooner." Finally, the *Nation* agreed with other publications that due to increased congressional and presidential influence and the fact that so many courts-martial occurred, and so few convictions resulted in significant punishment, it made sense that cadets, who were expected to uphold the values of the institution, were frustrated with such seeming inaction. Therefore, Congress "cannot restore discipline" to USMA; only by stopping the academy from being "part of the party 'spoils'" would discipline be restored.[93]

Throughout 1871, many newspapers continued to publish similar articles, either praising or denigrating Smith for his cadetship and expressing frustration that the War Department downgraded his courts-martial sentences to keep him as a cadet. As a result, some articles cynically noted "that all the white boys be turned out of the academy, and the institution be left solely to the colored cadet Smith."[94] Others blamed the overall discipline problems among all cadets on Smith's presence at the academy, implying these issues would not exist without the tension Smith's cadetship presented at USMA.[95] Some publications even asserted that if USMA "can not be better regulated than is indicated by the late transactions it had better, for the credit of the country, be shut up."[96] However, the Clearfield, Pennsylvania, *Raftsman's Journal* instead asserted that because of the mistreatment Smith experienced, the officers at USMA needed to be "capable of appreciating the demands of the times, and . . . have nerve enough to fairly protect all cadets, whether boys or youth, black or white."[97]

As Smith's first year at USMA ended, it became clear that some in the press sought to shift attention elsewhere while other journalists remained focused on integration at the academy. In some cases, that attention included taking jabs at Republican leadership who "ought to learn a lesson from the example at West Point, the nursery of loyalty" to stop attempting to encourage civil rights and "social equality" for Black Americans.[98] In contrast, the *Cairo Daily Bulletin* thought that Smith "has had his day" and should "stand aside" to provide others press space and public attention.[99] When fewer articles on Smith appeared between February and May 1871, the press started to lose interest, especially when his court-martial sentence would not be publicly announced.[100] When Smith was mentioned, it was frequently a short and insulting blurb, such as his heels being so long to explain why others stepped on them, and labeling him as violent, "ill-tempered, malicious, and a liar."[101] The *New York Times* claimed fewer articles on Smith were because the "troubles" associated with him were "more carefully shielded from the public eye."[102] Another explanation could be that because at least one more Black cadet had arrived at USMA, the press considered that the new and more interesting story.[103]

While the evidence chronicled here only spans the first year of Smith's cadetship, it is representative of how deep sectional tensions influenced white public opinion through press coverage of integration at USMA. The frustrations with the academy continued, as did the clear divisions among how white Americans viewed any Black cadet who attempted to enter USMA's hallowed halls. Additionally, it became clear that as the decade continued, the emancipationist legacy of the Civil War was already becoming a distant memory for whites. Instead, focusing on sectional tensions and reconciliation became central to how white public opinion reacted to integration at USMA. Because Smith dared to challenge racism and white supremacy at the academy, it fueled both southern and northern narratives on how and why Smith's cadetship might fail. Using his resistance to racism as evidence that segregation should be the norm everywhere, white southern newspapers emphasized how entrenched such practices were already becoming in the 1870s. Many white northern newspapers, though, offered ardent defenses of white officers stationed at USMA, attempting to absolve them of any wrongdoing. Articles written in this genre placed blame entirely on the academy as an institution and a select few cadets rather than acknowledging that widely held prejudices remained in the United States that most certainly influenced all who entered the academy's halls. Even so, some support still existed among white newspapers for Smith and integration at USMA, though their influence was muted by more broadly held racist perspectives. These three perspectives, evinced in the printed press, represented wider white public opinion in the first decade after the Civil War and also tell how sectional tensions permeated all parts of American society at the time, influencing how whites reacted to Black cadets at USMA.

Notes

1. "Colored Cadet," *Evening Star* (Washington, DC), March 8, 1870.
2. "Letters from the People—Letter from Salem, Massachusetts," *New Era* (Washington, DC), March 17, 1870.
3. "A Colored Cadet," *Prescott Journal* (WI), March 17, 1870; "Butler's Colored Cadet," *Tipton Advertiser* (IA), March 24, 1870.
4. "Colored Cadet at West Point," *Evening Star,* April 7, 1870. See also "Colored Cadet," *Staunton Spectator and General Advertiser* (VA), April 12, 1870; "Colored Cadet at West Point," *Wilmington Journal* (NC), April 15, 1870; "Colored Cadet at West Point," *Virginia Free Press* (Charlestown, WV), April 18, 1870; "Mississippi Legislature," *Weekly Clarion* (Jackson, MS), April 28, 1870.
5. "At the Capital [Condensed from the Daily Papers]," *New Era,* April 14, 1870.
6. "The 'Colored Cadet'—The Lesson of his Persecution," *Weekly Clarion,* February 9, 1871. While this phrasing appears in a Mississippi newspaper, denoting extensive derision at the idea of a Black cadet, the phrase ironically echoes the feelings of northerners as they began to reconsider USMA's attempts to rectify their role in commissioning so many officers who would betray their oaths to the nation in 1861. For more information about USMA's changing perceptions of its previous graduates and the new oath of office that resulted after the Civil War, see Ty Seidule, *Robert E. Lee and Me: A Southerner's Reckoning with the Myth of the Lost Cause* (New York: St. Martin's, 2021), 182–90.
7. For scholarship on the history of racism, news publications, shaping public opinion, and citizenship, see Juan González and Joseph Torres, *News for All the People: The Epic Story of Race and the American Media* (London: Verso, 2011); Natalia Molina, *How Race Is Made in America: Immigration, Citizenship, and the Historical Power of Racial Scripts* (Berkeley: University of California Press, 2014); Robert M. Entman and Andrew Rojecki, *The Black Image in the White Mind: Media and Race in America* (Chicago: University of Chicago Press, 2001); and Tyler Stovall, *White Freedom: The Racial History of an Idea* (Princeton, NJ: Princeton University Press, 2021).
8. "Mississippi News," *Memphis Daily Appeal,* April 24, 1870. See also "Items in Brief," *Nashville Union and American,* April 30, 1870; "Personal," *Portland Daily Press* (ME), May 17, 1870; and "Miscellaneous News Items," *Bloomfield Times* (New Bloomfield, PA), May 24, 1870.
9. "A Colored Cadet," *Cheyenne Daily Leader* (WY), May 16, 1870; "The First Colored Cadet," *Richmond Daily Dispatch* (VA), May 17, 1870; "Personal," *Portland Daily*

Press, May 17, 1870; "News of the Day," *Charleston Daily News* (SC), May 19, 1870; "Our Railroad," *Spirit of Jefferson* (Charlestown, WV), May 24, 1870; "Scraps and Facts," *Yorkville Enquirer* (SC), May 26, 1870.

10. "Political," *Memphis Daily Appeal,* May 20, 1870. See also "West Point Ku Klux," *Daily Phoenix* (Columbia, SC), May 25, 1870; "Appointment of a Colored Cadet," *Weekly Oskaloosa Herald* (IA), May 26, 1870.

11. "An Embryo Major-General," *Weekly Clarion,* May 26, 1870.

12. "The West Point Revels," *New York Sun,* May 26, 1870. See also "Washington," *New Orleans Republican,* May 27, 1870; "The Colored Cadet at West Point," *Chicago Tribune,* May 31, 1870; "The Colored Cadet Making His Bow to Col. Black, of the Regular Army," *Daily Phoenix,* May 31, 1870; "The West Point Revels," *Evening Argus* (Rock Island, IL), June 1, 1870; "Arrival of a Colored Cadet at West Point," *Daily Milwaukee News,* June 1, 1870; "West Point: Arrival of a Colored Cadet," *Opelousas Journal* (LA), June 4, 1870; "Reception of the Colored Cadet at West Point," *New Era,* June 9, 1870; "The West Point Revels," *Tarboro' Southerner* (NC), June 23, 1870; "The Colored Cadet at West Point," *Weekly Arizona Miner* (Prescott), June 25, 1870.

13. "West Point," *New York Times,* May 28, 1870. See also "The First Colored Cadet at West Point," *Portland Daily Press,* June 3, 1870; "The First Colored Cadet," *Tiffin Tribune* (OH), June 16, 1870.

14. "News of the Day," *New York Times,* May 28, 1870.

15. "The Colored Cadet," *Charleston Daily News,* May 27, 1870. This two-line reference is not a reprint from another newspaper. Instead, it appears to be a very succinct summation of a *New York Sun* article and highlights only the few items that white South Carolinians might be interested in: Howard's arrival, the commotion surrounding it, and the hotel's refusal to serve him.

16. "West Point," *Nashville Union and American,* May 28, 1870.

17. Untitled, *Alexandria Gazette and Virginia Advertiser,* May 30, 1870.

18. "General News," *Daily Kennebec Journal* (Augusta, ME), June 2, 1870.

19. "West Point: The Opening of the Season, Military and Otherwise," *New York Herald,* June 2, 1870.

20. "West Point," *New York Times,* June 10, 1870. See also "The Colored Cadets at West Point," *Daily National Republican* (Washington, DC), June 11, 1870; "West point— The Colored Applicants for Cadetship—False Reports Corrected—Generous Actions of the Cadets—'Social Equality,'" *Grant County Herald* (Lancaster, WI), June 21, 1870. While these articles clearly intended to begin shifting blame to institutions and to admonish both faculty and cadets for racism, the reality was quite different at the academy. Additionally, Smith noted a visit from a reporter in surviving documents, so it seems both newspapers engaged in false reporting when it suited their narrative. For more information, see Rory McGovern's essay in this volume, as well as Rory McGovern, Makonen Campbell, and Louisa Koebrich, "'I Hope to Have Justice Done Me or I Can't Get Along Here': James Webster Smith and West Point," *Journal of Military History* 87.4 (October 2023): 964–1003.

21. "West Point," *New York Times,* June 10, 1870. See also "The Colored Cadets at West Point," *Daily National Republican,* June 11, 1870.

22. "Tribulation Among the Cadets," *New Era,* June 9, 1870.

23. "The Colored Cadet Question," *Chicago Tribune,* June 11, 1870. See also "Local Items," *Daily Phoenix,* June 14, 1870; "News of the Day," *Charleston Daily News,* June 14, 1870; "Washington," *New York Times,* June 11, 1870; "Personal," *Chicago Tribune,* June 17, 1870.

24. "Colored Cadets," *Philadelphia Evening Telegraph,* June 13, 1870. See also "Local Items," *Daily Phoenix,* June 14, 1870; "News of the Day," *Charleston Daily News,* June 14, 1870; "Washington," *New York Times,* June 11, 1870; "Personal," *Chicago Tribune,* June 17, 1870.

25. "Local Items," *Daily Phoenix,* June 14, 1870. See also "All Sorts," *States and Union* (Ashland, OH), June 22, 1870.

26. "Proceedings of the Medical Board assembled at West Point, NY in compliance with Special Orders No. 194 + 201 dated War Department A.G. Office Washington May 19 & May 23 1870," and Special Report No. 15, June 5, 1870, James W. Smith Files, Roll 1, M1002, Selected Documents Relating to Blacks Nominated for Appointment to the United States Military Academy during the Nineteenth Century, 1870–1887, RG 404, USMA Special Collections and Archives, West Point.

27. "West Point," *New York Herald,* June 14, 1870. See also "The Colored Cadets," *Evening Argus,* June 17, 1870; "The Colored Candidates—The Graduating Class," *Chicago Tribune,* June 18, 1870; "The Colored Cadets at West Point," *Wilmington Journal,* June 24, 1870.

28. "The Colored Race in Politics," *New York Times,* June 14, 1870; "Washington," *New York Times,* June 11, 1870. See also "The Colored Cadets of West Point," *Daily National Republican,* June 27, 1870.

29. "West Point," *New York Herald,* June 15, 1870.

30. "The West Point Problem," *New York Sun,* June 17, 1870; "The Negro Cadets—Their Reception at West Point—The Charges of Maltreatment Reiterated," *Richmond Daily Dispatch,* June 20, 1870; "The Negro Cadets—Their Reception at West Point—The Charges of Mal-treatment Reiterated," *Wilmington Journal,* June 24, 1870. See also "The Colored Cadets," *New Orleans Republican,* June 19, 1870; "The Chinaman in Massachusetts—The Colored Boys at West Point," *Daily Phoenix,* June 23, 1870; "The Negro Boy from South Carolina," *Fairfield Herald* (Winnsboro, SC), July 13, 1870. See also "The West Point Colored Cadets," *New Orleans Republican,* June 30, 1870.

31. "The West Point Problem," *New York Sun,* June 17, 1870; "The Chinaman in Massachusetts—The Colored Boys at West Point," *Daily Phoenix,* June 23, 1870. See also "The Colored Cadets of West Point," *Daily National Republican,* June 27, 1870.

32. "Washington; West Point," *Chicago Tribune,* June 25, 1870. See also "The Colored Cadets of West Point," *Daily National Republican,* June 27, 1870; "Washington; Personal," *New York Herald,* June 7, 1870.

33. Untitled, *Raftsman's Journal* (Clearfield, PA), June 29, 1870.

34. "West Point," *Paterson Daily Press* (NJ), July 12, 1870.

35. "Currish West Pointers," *New York Dispatch,* July 10, 1870.

36. "The Colored Cadet," *Waterbury Daily American* (CT), July 9, 1870

37. "Cadet Appointees," *Evening Star,* July 1, 1870; "Personal and Literary," *Weekly Panola Star* (MS), July 2, 1870; "Miscellaneous Items," *Alexandria Gazette and Virginia Advertiser,* July 2, 1870; "News of the Day," *Charleston Daily News,* July 4, 1870; "Generalities," *Public Ledger* (Memphis), July 7, 1870.

38. Untitled, *Burlington Weekly Free Press* (VT), July 8, 1870. See also "News of the Day," *South Branch Intelligencer* (Romney, VA), July 8, 1870.

39. "The Colored Cadet," *New York Herald,* July 7, 1870; "The Colored Cadet Complains," *Evening Star,* July 8, 1870; "The Colored Cadet," *Vicksburg Weekly Herald* (MS), July 9 and 16, 1870; "The Colored Cadet," *Chicago Tribune,* July 9, 1870; "The Colored Cadet," *Charleston Daily News,* July 11, 1870; "West Point," *New York Tribune,* July 12, 1870; "West Point," *Paterson Daily Press,* July 12, 1870; "Disgraceful Treatment of the Colored Cadet," *New Era,* July 14, 1870; "The Colored Cadet," *Manitowoc Pilot* (WI), July 14, 1870; "The Colored Cadet—His Hardships at West Point," *Shenandoah Herald* (Woodstock, VA), July 14, 1870; "The Colored Cadets at West Point," *Orleans Independent Standard* (Irasburgh, VT), July 19, 1870; "A Colored Cadet at West Point," *Sumter Watchman* (Sumterville, SC), July 20, 1870.

40. "Personal," *Rutland Weekly Herald* (VT), July 7, 1870; untitled, *Alexandria Gazette and Virginia Advertiser,* July 9, 1870; "Currish West Pointers," *New York Dispatch,* July 10, 1870; "The Colored Cadet—His Hardships at West Point," *Tarboro' Southerner,* July 14, 1870.

41. "Ungentlemanly Conduct of the West Point Cadets," *New York Herald,* July 14, 1870. See also "Review of the Week," *Daily Evening Traveller* (Boston), July 16, 1870; untitled, *Bolivar Bulletin* (TN), July 23, 1870; "Review of the Week—Personal," *American Traveller* (Boston), July 23, 1870.

42. "The Colored Cadets at West Point," *Orleans Independent Standard,* July 19, 1870.

43. "West Point," *New York Tribune,* July 12, 1870; "West Point," *Paterson Daily Press,* July 12, 1870; "Letter from General Howard to Colored Cadet J. W. Smith," *Chicago Tribune,* July 14, 1870. See also untitled, *Alexandria Gazette and Virginia Advertiser,* July 13, 1870; "Letter from General Howard," *Lowell Daily Citizen and News* (MA), July 15, 1870; "The Colored Cadet," *Daily Phoenix,* July 17, 1870; untitled, *Stark County Democrat* (Canton, OH), July 20, 1870.

44. Untitled, *New York Tribune,* July 12, 1870.

45. Untitled, *Stark County Democrat,* July 20, 1870.

46. "The Colored Cadet from South Carolina," *Yorkville Enquirer,* July 14, 1870. See also "Mixed Public Schools," *Wheeling Daily Register* (WV), July 15, 1870.

47. "Colored Cadets at West Point," *Daily Evening Traveller,* July 13, 1870; "The Colored Cadet," *Evening Star,* July 13, 1870; "Congress Yesterday," *New York Herald,* July 13, 1870; "The Colored Cadet at West Point," *Chicago Tribune,* July 13, 1870; "Forty-First Congress—House," *Wheeling Daily Register,* July 13, 1870; "XLIst

Congress—Second Session—House," *Portland Daily Press,* July 13, 1870. See also untitled, *Alexandria Gazette and Virginia Advertiser,* July 11, 1870.

48. "Congress," *Philadelphia Evening Telegraph,* July 13, 1870; "The Colored Cadet," *Chicago Tribune,* July 14, 1870; "Congress," *Alexandria Gazette and Virginia Advertiser,* July 14, 1870; "Congressional Summary," *Evansville Journal* (IN), July 14, 1870; "Congressional," *Memphis Daily Appeal,* July 14, 1870; "XLIst Congress—Second Session," *Portland Daily Press,* July 14, 1870; "XLI. Congress," *Union and Journal* (Biddeford, ME), July 15, 1870; "Congressional Proceedings," *Staunton Spectator and General Advertiser,* July 19, 1870.

49. "The Colored Cadet at West Point," *New York Herald,* July 14, 1870; "Complaint of the Colored Cadet," *Chicago Tribune,* July 14, 1870; "Congress," *Alexandria Gazette and Virginia Advertiser,* July 14, 1870; "House," *Nashville Union and American,* July 14, 1870; "Washington News," *Richmond Daily Dispatch,* July 14, 1870; "Domestic News," *Daily Phoenix,* July 14, 1870; "Congressional Summary," *Evansville Journal,* July 14, 1870; "Washington," *Evening Argus,* July 14, 1870; "Congressional," *Columbian and Bloomsburg Democrat* (Bloomsburg, PA), July 15, 1870; "XLI. Congress," *Union and Journal,* July 15, 1870; "Congressional Proceedings," *Staunton Spectator and General Advertiser,* July 19, 1870; "Congressional," *Tiffin Tribune,* July 21, 1870; "Congressional," *Belmont Chronicle* (Clairsville, OH), July 21, 1870; "General News," *Iowa Voter* (Knoxville, IA), July 21, 1870; "Congressional," *Ottumwa Courier* (IA), July 21, 1870; "Congressional," *Columbian and Bloomsburg Democrat,* July 22, 1870.

50. "The Case of the Colored Cadet at West Point," *New York Herald,* July 15, 1870; "The Colored Cadet," *New York Times,* July 15, 1870; "Washington," *Chicago Tribune,* July 15, 1870; "News of the Day," *Alexandria Gazette and Virginia Advertiser,* July 15, 1870; "Telegraphic Items," *Portland Daily Press,* July 15, 1870; "Interesting Washington News," *Cheyenne Daily Leader,* July 15, 1870; "From New York," *Auburn Daily Bulletin* (NY), July 15, 1870; "The Case of Colored Cadet Smith," *Boston Herald,* July 15, 1870; "The Colored Cadet—Charges of Ill-Treatment to be Investigated," *Daily Evening Traveller,* July 15, 1870; "The Case of Cadet Smith," *Paterson Daily Press,* July 15, 1870; "The Colored Cadet," *Philadelphia Inquirer,* July 15, 1870; "The Colored Cadet at West Point," *Baltimore American and Commercial Advertiser,* July 15, 1870; "General News," *Daily Kennebec Journal,* July 18, 1870; "All Sorts," *Kenosha Telegraph* (WI), July 21, 1870; "The Colored Cadet," *Daily Alta California* (San Francisco), July 25, 1870; "The Colored Cadet at West Point," *Lowell Daily Citizen and News* (MA), July 25, 1870; "The Colored Cadet," *New York Herald,* August 13, 1870. Multiple newspapers listed all four, or mainly Wilson and Abbot as generals: during the Civil War they had been promoted to either brevet major general or brevet brigadier general.

51. "Troubles of a Colored Cadet," *Delaware Gazette* (Delaware, OH), July 22, 1870; "The Colored Cadet," *Chicago Tribune,* July 16, 1870; "The Colored Cadet at West Point," *Hartford Daily Courant* (CT), July 15, 1870; "The Colored Cadet at West Point," *Christian Register* (Boston), July 23, 1870.

52. "The Negro Cadet at West Point," *Wheeling Daily Register,* July 15, 1870. Of note, Wheeling resides in Ohio County, which in 1860–61 voted in favor of remaining in

the Union before war broke out. See also untitled, *Cambria Freeman* (Ebensburg, PA), July 28, 1870.

53. "The Colored Cadet," *Richmond Daily Dispatch,* July 18, 1870.

54. "Pot and Kettle," *Daily Missouri Republican* (St. Louis), July 19, 1870.

55. "The Tribulations of a Man and Brother among the Loyal Cadets at West Point— The Lesson," *Weekly Clarion,* July 28, 1870.

56. "The Colored Cadet," *New York Times,* July 20, 1870. See also "Washington Items," *Rutland Weekly Herald* (VT), July 21, 1870; "The Colored Cadet," *Daily Phoenix,* July 24, 1870.

57. "The Nigger in the West Point Woodpile," *Richmond Daily Dispatch,* July 25, 1870.

58. "Crowding Matters," *Cairo Daily Bulletin,* July 26, 1870.

59. "The Troubles Ahead," *Weekly Clarion,* August 4, 1870. See also "West Point," *Weekly Clarion,* August 25, 1870; "President Grant," *New York Herald,* August 29, 1870.

60. "The Colored Cadet Case," *New York Times,* August 13, 1870. See also "Washington," *Dubuque Daily Times* (IA), August 13, 1870; "The Colored Cadet," *New York Herald,* August 13, 1870; "News of the Day," *Alexandria Gazette and Virginia Advertiser,* August 13, 1870; "The Colored West Point Cadet Reprimanded," *Nashville Union and American,* August 13, 1870; "Washington Items," *Evansville Journal,* August 13, 1870; "The Colored Cadet Investigation," *Evening Star,* August 13, 1870; "Washington," *Memphis Daily Appeal,* August 13, 1870; "The Colored Cadet," *Portland Daily Press,* August 13, 1870; untitled, *Green-Mountain Freeman* (Montpelier, VT), August 17, 1870; "Miscellaneous News Items," *Prescott Journal* (WI), August 17, 1870.

61. "The Colored Cadet," *Chicago Tribune,* August 6, 1870. See also untitled, *Bolivar Bulletin,* August 6, 1870.

62. "The Colored Cadet on His Muscle," *New York Sun,* August 29, 1870. See also "General News," *Daily Kennebec Journal* August 31, 1870; "Colored Cadet on His Muscle," *Public Ledger,* September 5, 1870; untitled, *Vicksburg Weekly Herald,* September 10, 1870; "Knowledge & Goodnature," *Democratic Enquirer* (McArthur, OH), September 21, 1870.

63. "The Colored Cadet at West Point," *Evening Star,* August 31, 1870; "The Colored Cadet at West Point," *Alexandria Gazette and Virginia Advertiser,* September 1, 1870. See also "West Point," *New York Daily Tribune,* September 7, 1870.

64. "West Point," *New York Times,* October 21, 1870.

65. "The Colored Cadet at West Point, *Chicago Tribune,* October 22, 1870.

66. "The Colored Cadet at West Point," *Evening Star,* October 14, 1870; "Army Intelligence," *New York Herald,* October 14, 1870; "Washington," *New Orleans Republican,* October 14, 1870; "Special Court-martial," *Richmond Daily Dispatch,* October 14, 1870; "Washington Items," *Charleston Daily News,* October 14, 1870; "More Persecution of the Colored Cadet," *Nashville Union and American,* October 15, 1870; "Gossip and Gleanings," *Portland Daily Press,* October 17, 1870.

67. "West Point," *New York Times,* October 22, 1870. See also "West Point," *New York Herald,* October 22, 1870; "West Point," *Nashville Union and American,* October 22, 1870; "A Court Martial at West Point," *Portland Daily Press,* October 24, 1870.

68. "The West Point Court-Martial—More Testimony about the Colored Cadet," *New York Times*, October 23, 1870. See also "West Point. The Court Martial of the Colored Cadet," *Philadelphia Evening Telegraph*, October 24, 1870; "Military and Naval," *New York Tribune*, October 24, 1870.

69. "West Point," *New York Times*, October 26, 1870. See also "The Colored Cadet. His Trial by Court-Martial," *Evening Star*, October 27, 1870; "News of the Day," *Alexandria Gazette and Virginia Advertiser*, October 28, 1870; "General Press Dispatch," *New York Tribune*, October 29, 1870.

70. "Items of News," *Middletown Transcript* (DE), October 29, 1870.

71. "Colored Cadet Smith," *Chicago Tribune*, November 1, 1870; "End of the Colored Cadet," *Nashville Union and American*, November 1, 1870; "Telegraphic Summary," *Wheeling Daily Register*, November 1, 1870.

72. "The Court Martial of the Colored Cadet," *New York Herald*, November 16, 1870; "News of the Day," *Alexandria Gazette and Virginia Advertiser*, November 16, 1870.

73. "West Point," *New York Times*, November 17, 1870. See also "The Cadets Indignant," *New York Herald*, November 17, 1870; "Excitement at West Point—The Cadets Indignant at the Order Relieving Their Colored 'Brother in Arms' from Arrest," *Richmond Daily Dispatch*, November 19, 1870; "The Colored Cadet," *Charleston Daily News*, November 21, 1870; "The Colored Cadet," *Greenville Enterprise* (SC), November 23, 1870; "Indignant Cadets," *Public Ledger*, November 25, 1870.

74. "The Colored Cadet Case," *New York Times*, November 17, 1870.

75. Untitled, *Fayetteville Observer*, November 3, 1870.

76. Untitled, *Litchfield Enquirer* (CT), November 3, 1870.

77. "The Colored Cadet," *New Orleans Republican*, November 6, 1870.

78. "The Woes of West Point," *Charleston Daily News*, December 17, 1870. See also "Almost a Revolution," *Public Ledger*, December 19, 1870.

79. See Ulysses S. Grant Papers, 1819–1974, Library of Congress, Manuscript Division, Washington, DC; John Y. Simon, ed., *The Papers of Ulysses S. Grant* (Carbondale: Southern Illinois University Press, 2003).

80. "The Colored Cadet Again," *New York Herald*, December 9, 1870; "Preserving Discipline at West Point—Cadets Dismissed for Insubordination," *Wilmington Journal*, December 16, 1870; "The Colored West Point Cadet gain in Trouble," *New York Herald*, December 31, 1870; "The Colored Cadet Again in Trouble," *Nashville Union and American*, January 1, 1871; "The Colored West Point Cadet Again in Trouble," *Richmond Daily Dispatch*, January 3, 1871; "Case of the Colored Cadet," *Chicago Tribune*, January 6, 1871; "West Point," *Nashville Union and American*, January 6, 1871; "The Colored Cadet," *Wheeling Daily Register*, January 6, 1871; "That Irrepressible Colored Cadet," *Wheeling Daily Intelligencer*, January 6, 1871; "The West Point Negro," *Memphis Daily Appeal*, January 6, 1871.

81. "The Colored West Point Cadet Again in Trouble," *New York Herald*, December 31, 1870; "News of the Day," *Alexandria Gazette and Virginia Advertiser*, January 2, 1871. See also "The Colored West Point Cadet Again in Trouble," *Richmond Daily*

Dispatch, January 3, 1871; "The Woes of a Colored Cadet," *Charleston Daily News,* January 5, 1871; untitled, *Valley Virginian* (Staunton, VA), January 5, 1871; "The Woes of a Colored Cadet," *Sumter Watchman,* January 11, 1871.

82. Untitled, *Daily State Register* (Carson City, NV), January 3, 1871.

83. "The Colored Cadet in Trouble Again," *New York Herald,* January 4, 1871.

84. "West Point. Colored Cadet Court Martial," *Nashville Union and American,* January 5, 1871; "Telegraphic Summary," *Wheeling Daily Register,* January 5, 1871; "The Colored Cadet Again," *Memphis Daily Appeal,* January 5, 1871.

85. "General News," *Buchanan County Bulletin* (Independence, IA), January 6, 1871.

86. "The Colored Cadet," *New York Times,* January 5, 1871. See also "The Colored Cadet," *Chicago Tribune,* January 8, 1871. Other articles also noted a potential "conspiracy," but that was the concluding sentence, with no other indication that the paper or its correspondents held any sympathy toward Cadet Smith.

87. "West Point," *New York Times,* January 6, 1871.

88. "West Point," *New York Times,* January 7, 1871; "The Cadet Trial," *New York Times,* January 8, 1871; "West Point," *New York Times,* January 9, 1871; "West Point," *New York Times,* January 10, 1871; "Further Examination in the Case of Cadet Smith," *New York Times,* January 11, 1871; "West Point," *New York Times,* January 12, 1871; "West Point," *New York Times,* January 13, 1871.

89. "News by Latest Mails," *Portland Daily Press,* January 7, 1871; "West Point," *Nashville Union and American,* January 8, 1871; "That Colored Cadet" and "West Point Cadets in Trouble," *Memphis Daily Appeal,* January 8, 1871; "The Latest News. General and Domestic News" and "Case of the Colored Cadet," *Wheeling Daily Intelligencer,* January 9, 1871; "Washington News and Gossip," *Evening Star,* January 9, 1871; "The Colored Cadet Again in Trouble," *Camden Journal* (SC), January 12, 1871; "News of the Day," *Charleston Daily News,* January 13, 1871; "The Colored Cadet's Troubles," *Daily Phoenix,* January 14, 1871. For lengthier discussions of the court-martial proceedings, see "West Point Troubles," *Philadelphia Evening Telegraph,* January 11, 1871; "West Point," *New National Era* (Washington, DC), January 12, 1871; "What Will Be Done with Him? The Trial by Court Martial of the Colored Cadet," *New York Herald,* January 12, 1871; "West Point," *Philadelphia Evening Telegraph,* January 13, 1871.

90. "Ku-Klux at West Point," *New York Tribune,* January 10, 1871. See also "Mob Law at West Point," *Chicago Tribune,* January 13, 1871.

91. "That Colored Cadet Again," *New York Herald,* January 12, 1871.

92. "West Point. What a High Military Authority Has to Say about the Academy," *New York Herald,* January 13, 1871; "The West Point Colored Cadet," *New York Times,* January 14, 1871; untitled, *Alexandria Gazette and Virginia Advertiser,* January 14, 1871; "The Colored Cadet Case at West Point," *Wheeling Daily Intelligencer,* January 14, 1871; "Voice of the People. The Late Troubles at West Point," *New York Herald,* January15, 1871; "Colored Cadet," *Chicago Tribune,* January 18, 1871; "The Colored Cadet," *New York Times,* January 18, 1871; "The Colored Cadet," *Philadelphia Evening Telegraph,* January 20, 1871; "Congress Convulsed over the Colored

Cadet," *Nashville Union and American,* January 20, 1871; "Washington," *New York Times,* February 15, 1871; "The Difficulties at West Point," *New York Times,* January 15, 1871; "The West Point Troubles," *New York Times,* February 17, 1871.

93. "The West Point Troubles," *The Nation,* no. 293 (February 9, 1871): 84–85. See also untitled, *Wheeling Daily Register,* February 9, 1871.

94. Untitled, *Wheeling Daily Register,* January 20, 1871. See also untitled, *Green-Mountain Freeman,* January 25, 1871.

95. "The Dissensions in the West Point Academy," *New York Herald,* February 4, 1871; "The West Point Squabbles," *Philadelphia Evening Telegraph,* February 15, 1871. See also "The West Point Outrages" and "The West Point Investigation," *Chicago Tribune,* February 15, 1871; "Congressional. House," *Wheeling Daily Register,* February 15, 1871; "Outlived Its Usefulness," *Public Ledger,* February 15, 1871; "Congressional," *Memphis Daily Appeal,* February 15, 1871.

96. "West Point and Its Scandal," *Daily Telegraph* (Monroe, LA), February 13, 1871. See also "The West Point Outrages" and "The West Point Investigation," *Chicago Tribune,* February 15, 1871; "West Point Has Outlived Its Usefulness," *Nashville Union and American,* February 15, 1871; "Congressional. House," *Wheeling Daily Register,* February 15, 1871; "Outlived Its Usefulness," *Public Ledger,* February 15, 1871; "Congressional," *Memphis Daily Appeal,* February 15, 1871.

97. "Washington City Gossip," *Raftsman's Journal,* February 15, 1871.

98. "The Legislature," *Weekly Clarion,* March 2, 1871.

99. Untitled, *Cairo Daily Bulletin,* April 8, 1871.

100. "Items of News," *Democratic Advocate* (Westminster, MD), April 20, 1871; "West Point Troubles," *Philadelphia Evening Telegraph,* April 22, 1871; "News of the Day," *Charleston Daily News,* May 1, 1871; "Washington Items," *Yorkville Enquirer,* May 4, 1871.

101. "Scissorings," *State Rights Democrat* (Albany, OR), March 24, 1871; "News of the Day," *Charleston Daily News,* April 20, 1871; "Personal," *Cairo Daily Bulletin,* April 21, 1871.

102. "The Highlands. The Beauties of the Opening Spring Affairs at the West Point Academy and the Hotels," *New York Times,* April 14, 1871.

103. "Clippings," *Manitowoc Tribune* (WI), May 11, 1871; "West Point," *New York Times,* May 26, 1871. See also "Telegraphic Brevities," *Chicago Tribune,* May 27, 1871; untitled, *The Weekly Caucasian* (Lexington, MO), May 27, 1871; "Colored West Point Cadets," *Chicago Tribune,* May 29, 1871; "Miscellaneous Items," *Alexandria Gazette and Virginia Advertiser,* May 30, 1871; "News Brevities," *Richmond Palladium* (IN), June 3, 1871; untitled, *Portland Daily Press,* June 3, 1871; "The Military Academy," *New York Herald,* June 6, 1871.

"A BRIEF MOMENT IN THE SUN"

The Black Community's Response to a Failed Promise

Le'Trice D. Donaldson

*J*ubilant. *Stymied. DETERMINED.* These words encapsulate the sentiment of the African American population about their future during the Reconstruction era. Breaking the color barrier at almost any moment in American history can be taxing and traumatizing. When James Webster Smith emerged as the "chosen one" to integrate the United States Military Academy, he rode in on a wave of radical change that ultimately failed in providing a solid cadre of Black officers to command Black soldiers. The years following the Civil War offered a new life and meaning to the lives of the thousands of African American men who fought to suppress the Confederacy. Ronald Krebs accurately asserts, "Militaries are more than war-fighting machines: they are important sites of social and cultural power and contestation. More specifically, military institutions and service have, in the popular imagination, long been linked with citizenship and nationhood." The military service and performance of the African American community during the Civil War shattered the last remaining barrier to full citizenship access and rights for African American men. The Black community cultivated joy and racial pride in having Blackness represented in a US military uniform. The Black soldier served as a cornerstone for racial uplift in post-Reconstruction Jim Crow America.[1]

"The freedom of the Black race has been achieved by war," by blood, and by iron. As such, the expectations for Black men to be appointed and promoted from noncommissioned officers in the United States Colored Troops (USCT) to commissioned officers in the regular military were high. The

African American community overwhelmingly supported the appointment of Black cadets to West Point. They desperately wanted Black soldiers to be overseen by Black officers. Yet as the nineteenth century ended, it became painstakingly clear that the best route to getting Black officers in the field with Black soldiers would not be through *Old West Point*.[2]

A few years after World War I ended, a not-so-secret study was done by three white officers commissioned by Maj. Gen. H. E. Ely, Commandant of the Army War College. On November 10, 1925, Ely submitted this study to the War Department and recommended that this supposedly scientific analysis of the "Employment of Negro Manpower in War" be made the official policy of the War Department going forward as a means to address the "Negro Problem." In the thirty-three-page pseudoscientific report rooted in deeply ingrained paternalistic racism, the authors proclaimed, "The Negro does not perform his share of civil duties in time of peace in proportion to his populations. . . . Compared to the white man, he is admittedly of inferior mentality. He is inherently weak in character." The study states that white officers almost always commanded Black soldiers during the wars before the World War. When white leadership fails, it is dismissed as inefficient and unworthy of further analysis. "The Negro has run true to form. . . . He is exceptionally well fitted for Pioneer, Supply Trains, Labor Troops Remount Depots. . . . Experience has shown that cooks and waiters at officers' messes might well be colored." The opinions and observations in this report almost directly reflect the attitudes and policies of the regular army after the Civil War. The tone for the following century was set at the United States Military Academy under the presidential administration of Ulysses S. Grant.[3]

The Union forces at Fort Monroe in Virginia faced a peculiar dilemma in the spring of 1861: what to do with the looming runaway slave population. The commander, an attorney and notorious spoon thief from Massachusetts, Maj. Gen. Benjamin F. Butler, utilized this unintentional political savoir-faire at the start of the Civil War to implement psychological warfare against the Confederate insurrectionists and enact military emancipation for the unfree population. This, in turn, helped to trigger a chain reaction that would eventually lead to African Americans across the rebelling states freeing themselves and fighting for the boys in Union blue. It was Butler, in 1867, now a congressman, who first proposed admitting Black cadets to the "Nation's School." The desire to build institutions and bastions of education in their communities helped create both a launchpad and a safety net for the first Black cadets to attend the USMA. This essay will explore the tumultuous and ever-resolute support by the African American community of the trailblazers they hoped would lead their warriors into battle—and

the institutions, militias, and drill teams they created to train future Black
officers and military leaders.[4]

A Dream Deferred

Secretary of War Edwin Stanton sat unmoved as he read through the petition
and letters he received on behalf of Sgt. Maj. Christian Fleetwood. As the Civil
War drew to a close, it became clear that several officers' positions would be
available in the regular army. One of the white officers in Fleetwood's unit pe-
titioned the War Department to award him a commission. The sergeant major
was a Congressional Medal of Honor recipient; his citation reads: "Seized the
colors, after 2 color bearers had been shot down, and bore them nobly through
the fight." Fleetwood, the former editor of the *Lyceum Observer,* unceremoni-
ously received his medal rather anticlimactically in the mail several months
later. The desire to serve as a commissioned officer became Fleetwood's new
dream. Every officer in his regiment, all white men, submitted and signed the
petition to the War Department to have him commissioned an officer, a sure
sign of the respect felt by all who witnessed his gallantry. This petition focused
on Fleetwood's military service during the Civil War and sought to get him promoted to the rank of second lieutenant. The petition stated, "Sergeant Major
C. A. Fleetwood has proved himself as a gentleman and a soldier in every way
during daily intercourses extending over considerably more than a year." There
were several vacancies in the line officer corps. Still, Secretary of War Stanton,
unmoved by these ardent pleas, denied the request of the white officers because
"there was no law under which a colored man could be made an officer." At least
eighty commissioned Black officers fought in the Civil War. Yet none would be
commissioned into the regular army when the war ended.[5]

After this bitter disappointment at not earning a commission, Fleetwood
proclaimed, "No member of this regiment is considered deserving of a commission or if so cannot receive one . . . [and must continue] to act in a subordi-
nate capacity, with no hope of advancement or promotion." Yet Fleetwood's
military career did not end with Stanton's proclamation. He left the army in
1866 and took a nearly twenty-year hiatus from the uniform. But not long
after the war ended, the systematic erasure of African Americans' positive
contributions in the defeat of the Confederacy began, and it became clear
that something needed to be done. Writer, editor, and AME preacher Robert C. O. Benjamin astutely observed in 1896, "The pen of the historian has
been used to almost exclude any reference to the service of Negro soldiers
in the Union army. It is unknown that any Black men ever distinguished
themselves as soldiers." Benjamin had answered the call in 1870 when James

Webster Smith could enact and exercise his full rights as an American citizen by attending the US Military Academy to honor those who came before him and ensure they would not be forgotten. Meanwhile, a decade later, in 1880, after years of federal service, Fleetwood joined a growing movement to create state-funded Black militia units organized to be counterinsurgency units to combat white terrorist organizations such as the Ku Klux Klan and White Citizen Leagues.[6]

Fleetwood's shift to organizing the first African American National Guard Separate Battalion unit in Washington, DC, highlights a similar path that Smith would follow upon his eventual expulsion from West Point. With the denial of Fleetwood's petition to be granted an officer's commission, the hope became instead for a cadre of young Black men to exercise their full, first-class rights as citizens to graduate from the "Nation's School." The Black community found a way to adapt and carve new paths for the soldiers determined to become officers.

Elevation to a Higher Plain

"West Point, May 25, 1870,—the entire National Academy was almost breathless with excitement yesterday. The son of a colored American citizen arrived here in his new role as a military cadet. The white cadets seemed paralyzed.... 'Let's put the nigger in the river.' . . . Others talked of killing the Black boy outright." Upon arrival at West Point, James Webster Smith (fair-skinned) and Michael Howard (dark-skinned) entered an atmosphere thick with racial hatred. Even the academy professors boldly proclaimed, "It will be a long time before anyone belonging to the colored race can graduate at West Point." Yet Smith defiantly proceeded to push back and fight for his spot. With a sense of irony, he wrote a friend in 1870: "I forgot to tell you that out of 86 appointees, only 39 passed the examination. They had prepared it to fix the colored candidates, but it proved most disastrous to the whites." This simple statement in a private letter to his benefactor, which was eventually published, became a policy the Black community was determined to challenge and change. The fight to have Black soldiers commanded by Black officers was a rallying point for the African American community. In the minds of African Americans, military training and service served as confirmations of Black citizenship and manhood. Military training and educational opportunities validated their community's political and civil rights.[7]

Smith's and Michael Howard's admission into the academy inspired young Black men nationwide. Smith had been a student at Howard University before entering the academy. Webster Smith seemed to stand a better chance at survival because of his light complexion. Stephen B. Brague, an officer in the Ninth Army Corps, boldly proclaimed in his opening salvo in his 1863 Congressional report, "Notes Upon Colored Troops: "If you should ask me for the type of an admirable soldier, I would present you the Mulatto. It seems that he unites in himself, physically speaking, the perfection of both races." Capt. Rufus L. King, an academy tactical officer, went into excessive detail in describing Smith as "a tall, slim, loose-joined cadaverous party, with arms and legs of extraordinary length, and indescribable complexion, chalky white . . . the personification of gloom." King's description of Howard was far more basic and blunt: "A chuckly bucket-headed little darkie, whose great eyes wander." Yet Smith enjoyed no colorism advantage. Both Black men were initially surprised by the violent and hateful reception they received, but this did not stop them from pursuing their goal of becoming officers.[8]

James's father, Israel Smith, accompanied him to New York from South Carolina. Israel believed that the mistreatment and ungentlemanly behavior would cease once his son donned the uniform like the other cadets. He imparted sage advice to his son, which could account for how James Webster Smith survived four years of torment. Israel stated:

> You are elevated to a high position, and you must stand it like a man.
> Do not let them run you away, for then they will say, the "nigger" won't do.
> Show you[r] spunk and let them see that you will fight. When they find you are determined to stay, they will let you alone. You must not resign on any account, for it is just what the Democrats want.[9]

On June 28, 1870, James Webster Smith, encouraged by his father's words, officially entered the corps of cadets; Howard left the academy alone after failing the entrance exams.

Cadet Smith endured four years of malicious hazing and discrimination from his fellow cadets and professors. He did not always suffer alone or in silence. During his first year, Smith was court-martialed three times and found guilty in each case. Secretary of War William Belknap intervened after the recommended punishment for the second court-martial was dismissal from the academy. Belknap and the Republicans needed the support of the Black community; therefore, Smith was forced to repeat his first year. Henry Alonzo Napier joined Smith in 1871; Napier was a fellow Howard classmate. The two suffered together for one year, until Napier was dismissed for poor grades. Smith did surprisingly well with his grades, given the constant barrage of traumatic, racist treatment.

As One Star Rises, Another Fades

When Henry Ossian Flipper arrived at the academy in 1873, he expressed the same trepidation and determination Smith had in 1870. Flipper, the fourth Black cadet at West Point, said, "With a mind full of horrors of the treatment of all former cadets of color and dread of inevitable ostracism. I approached tremblingly, yet confidently."[10]

By 1874, Flipper and Smith were the only two Black cadets at the academy. James Webster Smith's failure to pass natural and experimental philosophy ended his tumultuous career at the academy. He took his case directly to Secretary of War William Belknap in Washington with Republican senator John J. Patterson of South Carolina. Nothing came of it; Belknap refused to restore Smith to the corps of cadets. Yet that did not silence Smith or remove him from maintaining a military role within the Black community. Smith accepted a position as commandant of cadets and professor of mathematics and military tactics at the South Carolina Agricultural and Mechanical Institute in 1875 and returned to the Hudson Valley in September 1875 to get married. According to the *Yorkville Enquirer,* he even secured West Point's military band for the occasion. Prominent Black leaders from across the country attended his wedding.[11]

During his years at West Point Smith was afforded neither support nor protection by the academy officials and staff or by Secretary of War Belknap. Not long after his dismissal in 1874, he described the lack of support and the abuse and persecution he faced from other cadets in a series of letters to Frederick Douglass's newspaper, the *New National Era and Citizen,* based in Washington, DC. Smith felt he owed it to his "race" to share his experience at West Point and expose his mistreatment. The *New National Era* happily provided a public platform for him. As early as 1871, after probably one of the most challenging years of Smith's life, the Black press had called for the abolishment of the academy. Asserting that "the Black press constructed Black soldiers as saviors and would-be martyrs for the benefit of the race," the *New National Era* fervently proclaimed the academy now "a dishonor to the nation; the men educated therein are more likely to prove a curse to the country in any great emergency."[12]

The African American community continued paying attention and applying pressure on the administrations of Grant, Hayes, and Arthur. The Black press and the support of powerful allies such as General O. O. Howard, Benjamin Butler, and Charles Sumner helped to push for a congressional investigation of the treatment Smith experienced. Smith's death in 1876 and the chapter about him in Flipper's 1878 autobiography demonstrated the value the African American community placed on his role as the first Black cadet admitted.

Henry O. Flipper's graduation from West Point on June 15, 1877, made him one of America's most famous Black men. There can be no doubt Flipper learned a great deal from watching Smith—so much so that Flipper dwarfed Smith's legacy overnight. He was now thrust into the national limelight. In the African American press, he emerged as a hero—at least until 1894, when the *Indianapolis Freeman* listed for sale a framed portrait of "Lieut. H. O. Flipper: Formerly of the U.S. Army, now of the Mexican." Otherwise, the African American press considered him on the same list as Toussaint L'Ouverture, George Washington Williams, Frederick Douglass, and Bishop Henry McNeal Turner. In many white newspapers, he emerged as a villain or cadet whose failure was guaranteed. The *Chicago Tribune,* for example, predicted that Flipper would never be allowed to graduate. "The prejudice of the regular army instructors against the colored race is insurmountable." The African American press (for example, the *Atlanta Herald*), described Flipper as above reproach: "Among colored men, we know of none more honorable or more deserving than Flipper. . . . Young Flipper is to make his mark as no other colored youth." Yet, Flipper's relationship with the other cadets at the academy became somewhat representative of how he treated the Black community for the remainder of his adult life: *distant.*[13]

Forever Climbing Onward

In the frigid winter of 1866, the officers and soldiers of the 62nd USCT founded a university in Missouri. The Lincoln Institute began as an interracial educational endeavor between white officers and Black soldiers. The men of the 62nd, mainly formerly enslaved men from Missouri, learned to read and write during their military service, and with these skills, their lives took on new meaning. "Whereas the freedom of the Black race has been achieved by war, and its education is the next necessity thereof, resolved, that we, the officers and enlisted men of the 62nd United States Colored Infantry, agree to give the sums annexed to our names to aid in the founding of an educational institution."[14] The men wanted a space for the newly emancipated to obtain a proper education. The excitement and determination to build, cultivate, and provide for their community at large is the mentality of racial uplift that spurred the creation of the only college in the United States ever founded by former soldiers and enslaved people.

The spirit and desire for military training in the African American community grew exponentially after the first Blacks were admitted to the academies. On its inaugural day in 1881, historian Marcus Cox astutely observed that the Tuskegee Institute's "first students marched in procession after inspection and were instructed in school policy and regulations. Booker T. Washington, the Institute's founder and principal, believed that military-type training was necessary to school discipline and instituted military ideals as an integral part of the curriculum." Hampton Institute and Wilberforce University each offered military training to their students. The Black community's demand and desire for military training opened them up to the possibility of military leadership, to ways to protect themselves and their communities, and for the young men to develop a truly "manly bearing."[15]

Therefore, despite the Black cadets' horrendous treatment and ruthless acts of violence against them, African Americans supported and fought to get their congressional representatives to nominate the best and the brightest young men from their communities to West Point. Yet the years following Flipper's graduation proved troublesome for Black cadets. In April 1880 Cadet Johnson C. Whittaker made national headlines after being accused of magically tying himself to his bed and slitting his own earlobes. Whittaker's case reached the desks of Secretary of War Robert Todd Lincoln and President Chester A. Arthur. During Whittaker's court-martial, Asa Bird Gardiner, the army's top trial attorney, kept referring to him as a member of an "inferior race" and framing Whittaker as a coward who would rather feign injury than take his exams. He was found guilty after two trials. However, in 1882, white allies such as Gen. Oliver O. Howard and former South Carolina governor Daniel Chamberlain convinced Lincoln to declare the court-martial invalid, thus sparing Whittaker a dishonorable discharge and imprisonment. But he did not get a happy ending. He was still dismissed from the academy for failing an exam in June 1880.[16]

Unlike the two other Black West Point graduates, John Hanks Alexander (1887) and Charles Young (1889)—not to mention Christian Fleetwood, Whittaker, and even Smith—Henry O. Flipper rejected the idea of ever working or teaching for an African American college or university or leading a Black militia unit. In his autobiography, he clarified that he had no intention of taking a position as a military tactics instructor. Flipper recalled, "When an army official suggested that [I] take an appointment at one of these schools, he said it would be best for me. I could not agree with him. Personally, I would rather remain with my company. I have no taste and no tact for teaching. I would decline any such appointment." This statement distinguishes Flipper from other Black military leaders and advocates of the late nineteenth century. Even his younger brother, Bishop Joseph Simeon Flipper, served as the commander of the all-Black Georgia Militia company the Thomasville Independents and became a dean and eventually the president of Morris Brown College. Bishop Flipper actively engaged in racial uplift and actively worked toward helping the African American community. There was even an all-Black Milita unit created in Petersburg, Virginia, called the Flipper Guard. Henry and Joseph both attended Atlanta University, yet after his departure from the army the former lieutenant decided to make his home in the rugged frontier of the southwest. Despite extensive traveling to rural Arkansas and Oklahoma, his brother always returned to Atlanta. He participated in the political movements and conversations surrounding the Black community and racial uplift.[17]

In fact, in 1902, when author and lecturer Daniel Wallace Culp published *Twentieth Century Negro Literature: Or a Cyclopedia of Thought on the Vital Topics Relating to the American Negro,* Joseph Flipper was among the one hundred contributors. The essays Culp collected were commissioned to address a specific question or topic. Joseph's piece addressed the question, "Is the Young Negro an Improvement, Morally, on His Father?" The other contributors to this discussion included George Washington Carver, Booker T. Washington, Timothy Thomas Fortune, Alice Dunbar (wife of Paul Lawrence Dunbar), and countless others from various professions. Culp posed thirty-eight questions, some dealing with education and racial uplift. Topics included "To what extent is the Negro pulpit uplifting the race?" "Will the education of the Negro solve the race problem?," and "The Negro as a writer; The Negro as an inventor." When Booker T. Washington needed an instructor for his military institute, he did not ask Flipper; the Tuskegee Wizard sought out Charles Young. Flipper criticized Black militias and regular troops on multiple occasions in the media. Given the views expressed in his autobiography and his disdain for volunteer militias, it is unsurprising that by the turn of the twentieth century, Flipper no longer held a holy place in the realm of "race leaders."[18]

On the evening of March 27, 1886, a troubled and sorrowful John Hanks Alexander wrote to his close friend John P. Green his reflection on the latest cowardly act of racial violence:

> My soul was deeply stirred to the very depths by the massacre of Colored people in Carrollton, Miss. . . . Because a colored man and his friends have the courage to stand up and resist and resent the impertinence of an insolent, overbearing white man, the latter's friends club together and, at the trial, open fire and coldly murder this colored man with 10 to 12 of his friends and wound several others. It was a damnable, perfidious, and cowardly act. Worthy of hellish imps as they are. . . . A man that can thus stand up for his right—his manhood—when low public sentiment all conspire to make him a cringing, cowardly, servile brute. . . . is more than a man, he is a hero. Would to God that such a spirit animated more of us.[19]

Alexander's lament reflected the pessimism and anger of the Black community toward these acts of racial violence. The weight of the race rested on Alexander's shoulders, and he understood that by simply finishing and graduating, his actions would be a source of inspiration. He suffered the same isolation and loneliness that Smith, Whittaker, and Flipper experienced but coped with the isolation by maintaining active communications with friends and family. The editor of the *Cleveland Gazette* kept a very close eye on the day-to-day happenings in Alexander's life. He routinely published updates for his readers because he knew the Black community cared. This is why the young cadet felt inspired to keep going in the face of hardship. "Alexander attempted to cope with his loneliness . . . by constantly reassuring himself that his success at the academy was of such importance to his race that it was worth any amount of sacrifice and hardship."[20]

After Charles Young graduated from West Point in 1889, it became increasingly clear that despite qualified candidates and applications (Benjamin Davis Sr., with the help of his mentor Charles Young, sought multiple ways to attend the academy), a chapter had closed on Blacks as cadets at USMA. In response, Wilberforce University, taking full advantage of the revised Morrill Land Grant Act in 1890, became the first Black college or university to establish a military science department. The Second Morrill Act was created to create and expand educational opportunities for people of color. Rather than land, money was designated for each state's land-grant institutions. The president of Wilberforce, Rev. Samuel T. Mitchell, successfully lobbied the two senators from Ohio to help pressure President Grover Cleveland and the War Department to not only appropriate funds for the new military science department but also to get one of the two Black field officers to oversee training the students. First Lt. Alexander, on orders from the War Department, arrived at Wilberforce as the new professor of military science and tactics in January 1894.[21]

"The Greatest Negro Soldier Alive"

The disappointment and disgust at not getting a commission after the Civil War took several years to subside in Christian Fleetwood. But he again was drawn to the uniform by creating the all-Black drill teams to compete locally and nationally. In 1880, Christian Fleetwood's Washington Cadet Corps joined the Butler Zouaves and the Stanton Guards—the first named for Gen. Benjamin Butler and the second for Edward Stanton. These two militia units served as territorial guards during the 1870s. With these three units combined, the Black community of the District had an entire battalion and created the

DC Guard. It was Christian Fleetwood, not Flipper, that the Black press in the late nineteenth century called "The Greatest Negro Soldier." Fleetwood set a standard for young Black cadets who aspired to leadership in the military. He embraced carrying the mantle for both his race and his country. Christian Fleetwood was more than just an officer; he was a race leader.[22]

In North Carolina, Charles Samuel Lafayette Alexander Taylor, a local fireman and political activist, organized the Charlotte Light Infantry in 1887. He served as the captain. Not long after the Charlotte Light Infantry was organized, Lt. John Alexander spent several days in the fall of 1891 training and inspecting the Black militia unit outside Raleigh, North Carolina. Taylor's unit would eventually be incorporated into the Third North Carolina Regiment of Volunteers, led by James Young, with C.S.L.A. Taylor as his second in command. This unit was one of three all-Black units—officers and soldiers alike—that served in the Spanish-American conflict that began in 1898.[23]

During the Spanish-American War, Flipper reemerged from private life to turn down support from the Black press and shoot down the ridiculous notion of him leading an all-Black militia unit. The *Cleveland Gazette* stated, "Ex-Lieut. Henry O. Flipper . . . does not use good judgment in issuing a letter requesting that newspapers and friends desist from any attempt to assist him in being reinstated in the army. The Afro-American press and his friends are the most powerful influences he can possibly bring to encourage and force them to do what has already been too long deferred." Yet Charles Young and Christian Fleetwood answered the call and led all-Black volunteer units as commanding officers. Some of the same officers Young trained at Wilberforce would join his staff with the Ohio volunteers.[24]

By the close of the nineteenth century, the African American community had long drifted away from the jubilance experienced at the end of the Civil War. The Black community's response to being stymied at the academy did not discourage their determination to have Black officers. African Americans understood, "more than any other factor, southern white supremacists feared that by fighting in the military, Black soldiers would see themselves as true American citizens worthy of social equality with whites and inevitably encourage other African Americans to think and act likewise." Wilberforce University regularly advertised young Black men in full uniform in Black newspapers and periodicals promoting their military science program. W. E. B. Du Bois wrote in 1935, "The slave went free, stood a brief moment in the sun, then moved back again toward slavery." It perfectly encapsulates the Black experience during Reconstruction and the Black experience at the United States Military Academy. Yet they persisted. When the Reserve Officer Training Corps program began, Black colleges and universities were among the first schools to accept cadets. The failed promise of honor and equality at *Old West Point* did not prevent or crush the dreams of a community determined to enact their rights as citizens and men.[25]

Notes

1. Ronald R. Krebs, *Fighting for Rights: Military Service and the Politics of Citizenship* (Ithaca, NY: Cornell University Press, 2006), 17. Terms such as "racial pride" and "racial uplift" are references specifically utilized by African American community leaders to counteract the nonstop barrage of racist stereotypes and racial violence inflicted on the Black community, and I use them as such. The uniqueness of the Black military experience lies in the understanding that one's service did not solely belong to oneself.

2. Albert Marshall, *A Soldier's Dream: A Centennial History of Lincoln University, 1866–1966* (Jefferson City, MO: Lincoln University, 1989), 3.

3. US Army War College, "Memorandum for the Chief of Staff regarding Employment of Negro Manpower in War, November 10, 1925," President's Official Files 4245-G: Office of Production Management: Commission on Fair Employment Practices: War Department, 1943, Archives of the Franklin Roosevelt Library, http://www.fdrlibrary.marist.edu.

4. Andrew Bledsoe, "Benjamin F. Butler and Military Emancipation," *Civil War Monitor* 11.4 (Winter 2021), https://www.civilwarmonitor.com/blog/benjamin-f-butler-and-military-emancipation.

5. Roger Cunningham, "'His Influence with the Colored People Is Marked': Christian Fleetwood's Quest for Command in the War with Spain and Its Aftermath," *Army History* 51 (Winter 2001): 20. James Harrison biographical sketch of Christian Fleetwood, 5–8, folder 1, Christian A. Fleetwood Papers, Schomburg Center for Research in Black Culture, New York Public Library. Christian Fleetwood served in the Fourth US Colored Infantry during the Civil War, enlisting in 1863 and serving until 1866. Due to Fleetwood's intelligence, he was quickly promoted to sergeant major and later served as a clerk for the War Department from 1881 until 1892. Fleetwood also wrote a pamphlet for the Negro Congress titled "The Negro as a Soldier," which was published by Howard University. In it, he traces the heroic deeds of Black soldiers starting with the American Revolution and ending with the Civil War.

6. Cunningham, "His Influence with the Colored People," 20; Richard Stillman, *Integration of the Negro in the U.S. Armed Forces* (New York: Frederick A. Praeger, 1968), 11; Robert C. O. Benjamin, *Light after Darkness: Being an Up-to-Date History of the American Negro* (Xenia, OH: Marshall & Beveridge, 1896), 18.

7. Henry O. Flipper, *The Colored Cadet at West Point: Autobiography of Lieut. Henry Ossian Flipper, First Graduate of Color from the U.S. Military Academy* (New York: Homer Lee & Co., 1878), 289, https://docsouth.unc.edu/neh/flipper/flipper.html#flipper289; Albert Williams, *Black Warriors: Unique Units and Individuals*

(Haverford, PA: Infinity Publishing, 2003), 18, 20; Marcus S. Cox, *Segregated Soldiers: Military Training at Historical Black Colleges in the Jim Crow South* (Baton Rouge: Louisiana State University Press, 2013), 2–3.

8. Thomas J. Fleming, *West Point: The Men and Times of the United States Military Academy* (New York: William Morrow, 1969), 219.

9. Flipper, *Colored Cadet at West Point,* 315–16.

10. Flipper, *Colored Cadet at West Point,* 29.

11. William S. McFeely, *Grant: A Biography* (New York: Norton, 1981), 376–77; Robert Eric Nedergaard, *Duty, Honor, Country, and Race: The Failure of Desegregation at West Point in the Nineteenth Century* (PhD diss., Arizona State University, Tempe, 2007), 288–90; *Yorkville Enquirer* (SC), September 9, 1875.

12. Chad Williams, *Torchbearers of Democracy: African American Soldiers in the World War I Era* (Chapel Hill: University of North Carolina Press, 2010), 48; William P. Vaughn, "West Point and the First Negro Cadet," *Military Affairs* 35.3 (October 1971): 101.

13. Flipper, *Colored Cadet at West Point,* 13–15; Le'Trice Donaldson, *Duty beyond the Battlefield: African American Soldiers Fight for Racial Uplift, Citizenship, and Manhood, 1870–1920* (Carbondale: Southern Illinois University Press, 2020), 114.

14. Marshall, *A Soldier's Dream,* 3. Lincoln University of Missouri was the first HBCU in the state. The men of the 62nd largely came from Missouri and were able to raise $6,000 to open its doors in 1866. Today, it currently has nearly two thousand students enrolled.

15. Maggi Morehouse, *The Routledge History of the American South* (New York: Routledge, 2017), 89.

16. To date, John F. Marszalek's *Court-Martial: A Black Man in America* (New York: Charles Scribner's Sons, 1972) remains the most definitive examination of the trials of Johnson C. Whittaker.

17. Marszalek, *Court-Martial,* 120. John Patrick Blair, *African American State Volunteers in the New South: Race, Masculinity, and the Militia in Georgia, Texas, and Virginia, 1871–1906* (College Station: Texas A&M University Press, 2023), 30, 45. The Flipper Guard was located in Petersburg, Virginia, and the Thomasville Independents was established in 1878, a year after Flipper graduated from West Point.

18. Andre E. Johnson, *No Future in This Country: The Prophetic Pessimism of Bishop Henry McNeal Turner* (Oxford: University Press of Mississippi, 2020), 3; D. W. Culp, ed., *Twentieth Century Negro Literature: Or a Cyclopedia of Thought on the Vital Topics Relating to the American Negro* (Naperville, IL: J. L. Nichols, 1902), 10–12, 257.

19. Willard B. Gatewood Jr., "John Hanks Alexander of Arkansas: Second Black Graduate of West Point," *Arkansas Historical Quarterly* 41.2 (Summer 1982): 122. Alexander had been a teacher not far from Carroll County. He knew the Brown brothers, the two Black men who were at the heart of the dispute with the white man, James Liddell. For more on the Carroll County Courthouse Massacre, please visit the Zinn

Education Project, "March 17, 1886: Carroll County Courthouse Massacre," https://www.zinnedproject.org/news/tdih/carroll-county-courthouse-massacre/.

20. Gatewood, "John Hanks Alexander," 120.

21. The 1890 Morrill Land Grant Act created the designation for Historically Black Colleges and Universities and nineteen new schools. For more information on the Second Morrill Land Grant Act of 1890 (Act of August 30, 1890, ch.841, 26 Stat. 417, 7 U.S.C. 322 et seq.), visit Black Past: African American History Timeline, https://www.blackpast.org/african-american-history/second-morrill-act-1890/; see also "Charles Young Teaching at Wilberforce," NPS.gov, Charles Young Buffalo Soldiers National Monument, https://www.nps.gov/articles/000/chyowilberforceuniver sity.htm.

22. Fleetwood Papers, Schomburg Center for Research in Black Culture, New York Public Library, folder 1, 8.

23. Gatewood, "John Hanks Alexander," 125.

24. *Cleveland Gazette,* May 28, 1898.

25. Williams, *Torchbearers of Democracy,* 46–48.

AFTERWORD

This volume, a collection of thematic essays, departs from earlier studies by addressing racial integration at West Point during Reconstruction from a variety of interrelated perspectives. In this way, we have attempted to place West Point's ultimately failed experience with racial integration during Reconstruction at the intersection of broader currents within the United States during that time. Like the war itself, Reconstruction pervaded nearly every aspect of American life, on social, political, and economic levels. The US Military Academy was not immune to issues affecting the nation writ large—the meaning of Black citizenship in the wake of emancipation foremost among them. Several important historical themes emerge from these essays, which not only offer new pathways to understand racial integration at West Point but also provide insights relevant to contemporary issues of race, access, and opportunity.

The military service and inarguable sacrifice of Black Americans during the war served as a very overt argument for their inclusion more generally in American public life. A strong record of valuable and often valiant service was an important threshold for later assertions of citizenship and more meaningful participation in the political process. This argument was not missed in the moment by either Black Americans or their supporters. A vocal minority of Union generals and politicians grasped this reality and promoted the use of Black troops in combat early in the war, and more came to share this sentiment after Black troops proved themselves at places like Port Hudson, Milliken's Bend, and Battery Wagner. Their wartime record stood on its own. Although some postwar antagonists attempted to credit entirely the leadership of white officers for Black troops' successes, such counterarguments were only the wails of those who knew the racial walls had been breached irreparably. It reasonably followed—at least for those who found credence in the promise of Abraham Lincoln's Emancipation Proclamation and the subsequent constitutional amendments—that young Black men would and should be offered the opportunity to prove themselves ready for an officer's commission. White and Black proponents alike pointed to the wartime record of the United States Colored Troops (USCT) as the beginning of a logical progression that would admit young Black men to West Point. If Black men could serve admirably in the ranks like white men, so too should they be extended the chance to earn an officer's commission. Accordingly, James Webster Smith's and Michael Howard's arrival at West Point in the summer of 1870 fulfilled a hope that began when Black troops flocked to the flag during the Civil War.

As this volume demonstrates, however, that same hope was not shared across white American society, nor even throughout the army or across the US Military Academy. Following the war, many if not most Americans were disturbed by the question of racial equality and the new freedoms ostensibly guaranteed to formerly enslaved Black Americans. This prompted reactionary responses to the integration of West Point among white cadets, faculty, army officers, and broad swaths of the white American public. While a valiant wartime record opened the doors of West Point for young Black men aspiring to become army officers, it ultimately was not enough to secure their access, foster their success, and keep those doors open for long. Smith's hardships, and the similar experiences of those Black men who followed him to West Point, are sad evidence of this travesty.

Another central thread in *Race, Politics, and Reconstruction* is the realization that the appointment of Black men to West Point took on a much larger meaning in American society given the context of Reconstruction. This assertion may seem to point out the obvious, but the essays in this volume give integration at West Point during Reconstruction a more nuanced and complete expression in the historical record. The social upheaval that characterized Reconstruction was no less present in the attempt to racially integrate West Point. It was more than a question of whether Black cadets could succeed or whether they had the necessary preparation to thrive as cadets or if white troops and white officers would accept them into the US Army. It also was a question of how Black accession to West Point reflected the broader political effort to reconstruct the American South, and by extension the country more generally, as a nation defined by equal opportunity under the law—a question that at the time remained still unanswered.

Like so many issues written against the backdrop of Reconstruction's deeply racialized partisan politics, an integrated West Point meant a victory for Republicans over Democratic opposition. The very existence of Black cadets at West Point carried much more significant political ramifications than even the question of race alone might have implied. It was a battlefield in the greater partisan war for control of the country—perhaps even an extension of the Civil War itself, not unlike how Southern "Redeemers" later sought to reverse the war's outcome. James Webster Smith and those who came after him could be forgiven for thinking they carried the weight of the world on their shoulders. Their daily existence as cadets and the responses of various public constituencies transcended their lessons in descriptive geometry, chemistry, or philosophy. Their dreams of an army officer's commission carried significance beyond questions of access. Instead, they spoke to the very question of who would guide and control the country's institutions. The Civil War's end and the Thirteenth Amendment settled the questions of secession and slavery in the United States. But the war continued in altered form with two political parties battling over racialized politics taking the place of armies fighting in the field. Thus, West Point's Black cadets were cast in a light widely perceived as inherently partisan.

Moreover, the effort to integrate the corps of cadets became mired in the related argument over the extent and degree of federal authority, and thus civil authority, over the army. That this question might have been part of resistance to a legal action at a decidedly federal institution chartered for a clear national purpose seems ironic now. But at the time, many Americans (most Democrats and even some moderate Republicans) viewed the Grant administration's Reconstruction policies, and thus martial governance in the South, as a stretch of federal authority with autocratic implications. White cadets and faculty alike resented Black cadets, whom they saw as a proximate reminder that "their" academy had become a partisan tool and its integration an expression of Grant's will but not their own or that of the people. Given the academy's obvious federal purpose, the insult they claimed at having to conform to Grant's policy of integration is curious indeed. But at that time the regular army's sense of subordination to civil authority was not nearly so well defined as it would later become in the twentieth century. When reactionaries at the academy and within the wider army expressed hatred for Black cadets, they did so not only in racial and racist terms but also in terms of professional autonomy. White cadets' outright rejection of the entire premise of integration was couched in terms of disgust for what they believed were its baneful politics.

Individually and collectively, the several essays presented here provide an anatomy of failure. The story of integration at West Point during Reconstruction is a story of lost opportunity, or, perhaps more accurately, an opportunity squandered with shocking degrees of forethought and deliberation. But it was a failure owned by many, both within and outside the academy. West Point's failure was the army's failure; and the army's failure was the country's. In this conclusion lies the singular critique that is this book's coda: in both its promise and its failure, integration at West Point was a microcosm of Reconstruction itself. To attempt to understand otherwise the racist reception and resistance that greeted Smith, Napier, Gibbs, Flipper, Williams, and Whittaker is to approach the problem without historical perspective. We hope this volume offers exactly that—a more complete understanding of a particularly complex story that abounds with lessons and insights on how societies and their institutions succeed and fail as they navigate the challenges of cultural change.

BIBLIOGRAPHIC ESSAY

This essay offers a brief summary of many of the important resources that help create an understanding of the US Military Academy's place in American history, particularly during the Reconstruction era. The works presented here are only a starting point. We hope that those interested in the subject more broadly or who pursue more specific research associated with West Point during the period may also find value in this discussion as a foundation for further reading. The literature presented focuses on the experiences of the first young Black men who arrived there to face the challenges of becoming a cadet amid the racism that was endemic both at West Point and within American society. The subject of this book lies at the intersection of four distinct bodies of historical scholarship: Reconstruction, the United States Army, West Point generally, and integration at West Point specifically. Salient works in each of these areas are presented as resources for further reading.

Reconstruction

Scholarship on the Reconstruction era is lively and vast. Although the broader subject is beyond the focused scope of this book's analysis of race and politics at West Point during the same period, excellent starting points include Eric Foner's *Reconstruction: America's Unfinished Revolution, 1863–1877,* updated ed. (New York: HarperCollins, 2014). Alongside Joshua Brown, Foner also published a collection of essays that relate Reconstruction's social-political themes to contemporary events and practices: *Forever Free: The Story of Emancipation and Proclamation* (New York: Knopf, 2005). More recent scholarship that has come to define the field includes Heather Cox Richardson, *The Death of Reconstruction: Race, Labor, and Politics in the Post–Civil War North, 1865–1901* (Cambridge, MA: Harvard University Press, 2004); Michael W. Fitzgerald, *Splendid Failure: Postwar Reconstruction in the American South* (Chicago: Ivan R. Dee, 2007); and Mark Wahlgren Summers, *The Ordeal of the Reunion: A New History of Reconstruction* (Chapel Hill: University of North Carolina Press, 2014). Brooks D. Simpson focuses more specifically on the role of the executive office in Reconstruction politics in *The Reconstruction Presidents* (Lawrence: University Press of Kansas, 1998). Most recently, Adam Domby and Simon Lewis have published a valuable collection of essays in *Freedoms Gained and Lost: Reconstruction and its Meanings 150 Years Later* (New York: Fordham University Press, 2022), which examines a diverse array of Reconstruction's themes, meanings, and legacies and offers an excellent starting point for students embarking on a study of the era.

A similarly broad body of work examines intersections between race, politics, and society in late nineteenth-century American society. Steven Hahn's *A Nation under Our Feet: Black Political Struggles in the Rural South, from Slavery to the Great Migration* (Cambridge, MA: Harvard University Press, 2003) continues to define the field. Hahn argues that Black political activity throughout the late nineteenth century was driven by a shared vision of autonomy and collective self-defense that grew from—in part—the experiences of wartime military service, postwar promises, and postwar backlash. The culture of racial prejudice that was undeniably a part of the post–Civil War United States is well analyzed in Leslie H. Fishel's essay, "The African-American Experience," in *The Gilded Age: Essays on the Origins of Modern America* (Wilmington, DE: Scholarly Resources, 1996). The struggle with race as a lasting social divide is addressed by John Hope Franklin and Alfred A. Moss Jr. in *From Slavery to Freedom: A History of African Americans,* 8th ed. (New York: Knopf, 2000). In the seminal *Educational Reconstruction: African American Schools in the Urban South, 1865–1890* (New York: Fordham University Press, 2016), Hilary Green offers excellent analysis of efforts to bridge that divide through African American schools in Richmond, Virginia, and Mobile, Alabama. On the history of racism in published media and how it has deeply influenced American public perspectives, see Juan González and Joseph Torres, *News for All the People: The Epic Story of Race and the American Media* (London: Verso Books, 2011). Importantly, González and Torres bring to light the roles played by alternative newspapers in a deliberate effort to overcome racially charged conventional media—a story that also is taken up in this volume's essays in the study of race during the Reconstruction period.

The US Army

General histories of the United States Army are legion, but two works come to mind as most important and relevant toward this topic. Edward M. Coffman's *The Old Army: A Portrait of the American Army in Peacetime, 1784–1898* (New York: Oxford University Press, 1988) remains the most thorough and astute social history of the US Army during the long nineteenth century.

More recently, Robert Wooster's excellent *The United States Army and the Making of America: From Confederation to Empire, 1775–1903* (Lawrence: University Press of Kansas, 2021) provides valuable insights into the complex and symbiotic relationship between nineteenth-century American politics and the army.

The US Army's role in Reconstruction, militarily and as an instrument of political purpose, is comparatively understudied. Scholars tend to focus more on the Freedman's Bureau than on the army itself in the Reconstruction era. Older and reliable starting points are Joseph Dawson, *Army Generals and Reconstruction* (Baton Rouge: Louisiana State University Press, 1982); Martin E. Mantell, *Johnson, Grant, and the Politics of Reconstruction* (New York: Columbia University Press, 1973); and James E. Sefton, *The United States Army and Reconstruction, 1865–1877* (Baton Rouge: Louisiana State University Press, 1967). But Gregory P. Downs's more recent and thoroughly researched *After Appomattox: Military Occupation and the Ends of War* (Cambridge, MA: Harvard University Press, 2015) stands above the rest.

Long neglected in American military history, an impressive and growing body of literature has shed new light on Black military service during and after the Civil War. This is an essential topic for anyone researching integration at West Point during Reconstruction, with clear and recognizable influences. The war brought Black military service into the national conversation and provided the examples and role models foremost in the minds of young Black men seeking appointments to West Point during Reconstruction. Dudley Taylor Cornish's classic history, groundbreaking at the time, is an excellent starting point: *The Sable Arm: Black Troops in the Union Army, 1861–1865* (1956; reprint, with a foreword by Herman Hattaway, Lawrence: University of Kansas Press, 1987), as is Joseph Glatthaar's *Forged in Battle: The Civil War Alliance of Black Soldiers and White Officers* (New York: Free Press, 1990), and William A. Dobak's more recent *Freedom by the Sword: The U.S. Colored Troops, 1862–1867* (Washington, DC: Center for Military History, 2011). In *Fighting for Citizenship: Black Northerners and the Debate over Military Service in the Civil War* (Chapel Hill: University of North Carolina Press, 2020), Brian Taylor shows how Black Americans used debates about and the experience of military service to forge new definitions of citizenship and nationhood. Holly A. Pinheiro Jr. provides a superb analysis of Black service in the Civil War as a means of claiming racial justice for soldiers, veterans, their families, and their communities in *The Families' Civil War: Black Soldiers and the Fight for Racial Justice* (Athens: University of Georgia Press, 2022). For fresh and penetrating insights into the motivations informing Black service after the war, turn to Le'Trice Donaldson's *Duty beyond the Battlefield: African American Soldiers*

Fight for Racial Uplift, Citizenship, and Manhood, 1870–1920 (Carbondale: Southern Illinois University Press, 2020).

African American service was one element within broader currents of change within the army. Changes elicited a wide variety of opinions from the officer corps. Researchers wanting to delve into how officers thought about the army, its institutions, and the world around them in the late nineteenth century should turn to contemporary military journals. At the time, professional journals increasingly became the touchstone for anyone who wished to claim expertise or specialized knowledge in a particular subject area. Many members of the army's officer corps became avid participants in this practice, and contemporary military journals offered a wide variety of thoughtful arguments on the day's issues. Articles ran a spectrum from evolving field hardware and newly invented weapons to contemporary events in Europe and critical discussion of the army's bitter experience in the American West, shedding valuable insight into officers' perspectives. *Journal of the Military Service Institute of the United States* (*JMSIUS*) and *Army and Navy Journal* (*ANJ*) both published robust discussions in short opinion pieces as well as longer topical articles and should be consulted as part of any serious research.

West Point and Integration at West Point

Examples of nineteenth-century military journal articles specific to the West Point experience are George L. Andrews, "The Military Academy and Its Requirements," *JMSIUS* 4.14 (1883): 112–46; and James B. Fry's "Admission to the Military Academy," *JMSIUS* 4.14 (1883): 101–11. Similarly, with broader address, officers argued West Point's professionalizing role and its value to American society. For instructive examples, see "Military Education," *ANJ* 3.28 (March 3, 1866) and "West Point Training," *ANJ* 3.29 (March 10, 1866). Peter S. Michie, a permanent member of the faculty in physical sciences and an energetic champion for the academy's pedagogical traditions, promoted "Educational Methods at West Point" in *Educational Review* 4 (November 1892): 350–65. Michie offered an address at Union College in 1895 that celebrated West Point as a bulwark of expertise that eventually carried the North to victory: see "A Personal Experience of the Influence of West Point Education in the Training of an Army Officer during the Time of War," handwritten draft, in Michie Papers, CU1996, US Military Academy Library Special Collections and Archives. Michie's words offer deep insight into the faculty perspective of the day—perhaps one of the greatest influences on cadets, given the faculty's pervasive dominance over the academy's austere environment.

Just as contemporary commentators thought and wrote about West Point, so too have American military historians. Stephen Ambrose, *Duty, Honor, Country: A History of West Point* (Baltimore: Johns Hopkins Press, 1966); Thomas Fleming, *West Point: The Men and Times of the United States Military Academy* (New York: William Morrow, 1969); and George Pappas, *To the Point: The United States Military Academy, 1802–1902* (Westport, CT: Praeger, 1993) all provide important though often uncritical histories of West Point. Pappas's *To the Point* is especially valuable for the primary excerpts he included alongside many historic photographs. Lance Betros's *Carved from Granite: West Point since 1902* (College Station: Texas A&M University Press, 2012) provides a more nuanced analysis, but its history of the academy is limited to the twentieth century. Theodore Crackel's *West Point: A Bicentennial History* (Lawrence: University Press of Kansas, 2002) offers the most recent and most balanced institutional history of West Point from its origins to the present day.

Since its establishment in 1802, West Point assumed a cultural significance that transcended its more utilitarian purpose of preparing officers to lead military operations and undertake America's engineering works. As an instrument of political patronage through the congressional nomination system, it understandably attracted interest that went beyond its daily lessons, traditions, and curriculum. Several narrative histories offer useful understanding of West Point's place in American history, both as a national resource and as a catalyst for contention. Among them are James L. Morrison's *The Best School in the World: West Point, the Pre–Civil War Years, 1833–1866* (Kent, OH: Kent State University Press, 1986). Morrison's work, an adaptation of his extensive dissertation research, provides a critical survey of the academy's growing influence on American society during the period following Sylvanus Thayer's departure. Morrison places this influence into the context of episodic tension with Jacksonian democrats, who attempted to paint the academy as a bastion of elitism, antithetical to America's preferred self-image.

Such criticism was amplified during and after the Civil War. On controversy surrounding wartime West Point as a national academy of claimed martial expertise and patriotic zeal, see T. Harry Williams, "The Attack upon West Point during the Civil War," *Mississippi Valley Historical Review* 25.4 (1939): 491–504; Alan Aimone and Barbara Aimone, "Much to Sadden and Little to Cheer: The Civil War Years at West Point," *Blue and Gray Magazine* (December 1991); and Lori A. Lisowski, "The Future of West Point: Senate Debates on the Military Academy during the Civil War," *Civil War History* 34.1 (1988): 5–21. These histories present West Point as a lightning rod for public discourse on the proper role of iconic institutions in a nation that celebrated its meritocratic identity. The resignation of so many graduates from US Army rolls as they took up rebellion in the Confederate army energized such criticism. This was especially true during the war's first years, when the Union suffered a string of high-profile battlefield defeats, too often at the hands of West Point graduates who had resigned to cast their lots with the Confederacy. Such criticism has been well covered by historians. It features prominently in all the major institutional histories of West Point. It is also the focus of James L. Morrison's "The Struggle between Sectionalism and Nationalism at Ante-Bellum West Point, 1830–1861" in *Civil War History* 19.2 (1973).

Less well covered is the first period of racial integration at West Point from 1870–89. This volume is, to our knowledge, the first that attempts a systematic examination of racial integration at West Point during the Reconstruction era. Tom Carhart's recently self-published *Barricades: The First African-American West Point Cadets and Their Constant Struggle for Survival* (Xlibris, 2020) is more biographical in nature; most of its chapters offer insight into the background and experiences of Black cadets of the period. Institutional histories of West Point tend to acknowledge and summarize the deeply contested character of integration at the academy but lack any particular depth of analysis covering the facts and process of integration or the people involved. One exception is Walter Scott Dillard's unpublished PhD dissertation, "The United States Military Academy, 1865–1900: The Uncertain Years" (University of Washington, 1972), in which a chapter titled "The Black Cadets: Dilemma, Agony, and Failure" presents what is likely the best critical narrative to date.

Biographical works have partially filled the gap. Some of the notable analyses of aspects of integration at West Point in the late nineteenth century appear in works devoted to individual cadets, officers, and political leaders who played roles large and small in the larger narrative of integration at West Point. A portion of William S. McFeely's *Grant: A Biography* (New York: W.W. Norton, 1981) examines James W. Smith's admission and experience at USMA from the administration's perspective. Smith and Henry O. Flipper also feature prominently in a section within chapter 5 of David J. Fitzpatrick's excellent *Emory Upton: Misunderstood Reformer* (Norman: University of Oklahoma Press, 2017). Johnson C. Whittaker's case receives critical analysis in chapter 11 ("The Mistake of My Life") of Donald B. Connelly's *John M. Schofield and the Politics of Generalship* (Chapel Hill: University of North Carolina Press, 2006). These works, however, remain focused on white figures who played a part in the attempt to integrate West Point during and after Reconstruction, a circumstance that imposes natural limits on the scope of their analyses of integration.

Biographical works of Black cadets of the late nineteenth century are rich in both detail and analysis but skew toward the exceptional cases of Henry O. Flipper, Johnson C. Whittaker, and Charles Young. Flipper and Young are exceptional because they graduated, unlike nine of twelve Black cadets admitted between 1870 and 1889. John Hanks Alexander also graduated, but his tragic, early death left a sparse historical record. Whittaker is exceptional because of the spectacular nature of his assault and subsequent court-martial. Flipper's own account is a wonderful starting point: *The Colored Cadet at West Point. The Autobiography of Lieut. Henry Ossian Flipper, U.S.A., First Graduate of Color from the U.S. Military Academy* (New York: Homer Lee & Co., 1878) is also available in several reprinted editions. Flipper's later court-martial on charges stemming from missing funds at Fort Davis, Texas, is covered by James M. Robinson, *The Court-Martial of Lieutenant Henry Flipper* (El Paso: Texas Western Press, 1994). Flipper's memoir of his time in the West as a young army officer is also available as *Frontiersman: The Memoirs of Henry O. Flipper*, edited by Theodore D. Harris (Fort Worth: Texas Christian University Press, 1997). John F. Marszalek details the experiences and eventual court-martial of Cadet Johnson Chesnut Whittaker, a young Black man who entered West Point in 1876, in *Court-Martial: A Black Man in America* (New York: Charles Scribner's Sons, 1972). Marszalek's valuable study places Whittaker's trial and his experience at West Point generally into the backdrop of Gilded Age America's social fabric. It was republished by Macmillan's Collier Books as *Assault at West Point: The Court-Martial of Johnson Whittaker* (New York: 1994). Charles Young benefitted from an excellent biographer in Brian G. Shellum, whose *Black Cadet in a White Bastion: Charles Young at West Point* (Lincoln, NE: Bison Books, 2006) offers a very readable analysis of Young's experience at the academy, and *Black Officer in a Buffalo Soldier Regiment: The Military Career of Charles Young* (Lincoln, NE: Bison Books, 2010) narrated Young's subsequent lengthy career. More recently, Le'Trice Donaldson's *Duty beyond the Battlefield*—already mentioned above—includes one chapter each on Henry O. Flipper and Charles Young, offering valuable analyses of their motivations and experiences. Most recently, James W. Smith received his first in-depth scholarly attention in the article "'I Hope to Have Justice Done Me or I Can't Get Along Here': James Webster Smith and West Point," featured in the October 2023 issue of *Journal of Military History* and written by three of this volume's contributors.

It is our hope that the present volume becomes as essential a source to those researching the first attempt to integrate West Point in the late nineteenth century as those we have identified above. We earnestly believe that no other work thus far has examined the issue so comprehensively and holistically. But as any experienced historian knows, no volume ever constitutes the last word. We look forward to other scholars building upon or modifying the analysis offered in these pages—the topic certainly remains relevant and warrants further address.

CONTRIBUTORS

Several contributors to this volume are currently serving or employed within the Department of Defense in various military or civilian capacities. This volume is not the product of officials of the Department of Defense acting in official capacities. The views and opinions expressed throughout this volume are the contributors' own and do not reflect official stances or positions of the United States Military Academy, the United States Marine Corps, the United States Army, or the Department of Defense.

JONATHAN D. BRATTEN is a historian of the American colonial era and World War I as well as an officer in the Army National Guard. He holds BA and MA degrees in history, having served as a military history instructor for both University of Maine ROTC and the US Military Academy at West Point. He is the author of *To the Last Man: A National Guard Regiment in the Great War, 1917–1919,* which received the 2020 Army Historical Foundation Distinguished Writing Award. In 2016, Bratten served as a historical on-screen adviser for the Smithsonian Channel documentary *Americans Underground: Secret Cities of World War One.* He has written for *Army History,* the *Washington Post,* and the *New York Times.* Bratten contributed chapters to the anthologies *Strategy Strikes Back: How Star Wars Explains Modern Military Conflict* and *Armies in Retreat: Chaos, Cohesion, and Consequences.*

MAKONEN CAMPBELL, a native of Charlotte, North Carolina, enlisted in the US Army in 1994 and was commissioned in 2009. After almost thirty years of active-duty service, he retired as a professor of history at the United States Military Academy in the rank of major. He holds a BA in history from the University of Maryland University College, an MS in Organizational Leadership from Columbus State University, and an MA in History from Clemson University. He is currently pursuing a PhD at Texas A&M University. His primary research focus is on civil rights and urban renewal in the South, though he has previously written on integration at West Point. He is a coauthor of "'I Hope to Have Justice Done Me or I Can't Get Along Here': James Webster Smith and West Point," in the *Journal of Military History.*

ADAM H. DOMBY is an Associate Professor of history at Auburn University. An expert on the Civil War and Reconstruction, he is the author of *The False Cause: Fraud, Fabrication, and White Supremacy in Confederate Memory*. He coedited with Simon Lewis *Freedoms Gained and Lost: Reconstruction and Its Meanings 150 Years Later*. An award-winning historian, his writing has appeared in the *Journal of Southern History* and *Civil War History* among other venues. He received his MA and PhD from the University of North Carolina at Chapel Hill and his BA from Yale University.

LE'TRICE D. DONALDSON is Assistant Professor of History at Auburn University. Dr. Donaldson specializes in late nineteenth- and early twentieth-century African American military history, the Gilded Age, World War I, and gender history. She works at the intersection of race, gender, military service, and the long civil rights movement. She is the author of *Duty beyond the Battlefield: African American Soldiers Fight for Racial Uplift, Citizenship, and Manhood, 1870–1920* and *A Voyage through the African American Experience*. She is also an editor for a new book series, The Black Soldier in War and Society: New Narratives and Critical Perspectives, with the University of Virginia Press. Her current book projects are *Remember Me: The Life of Eugene Bullard, the First African American Fighter Pilot*, and *Race Prophets: A History of the Army's Black Chaplains*. Dr. Donaldson is the founder and president of the Society for Black Military Studies, currently serves on the executive board of the Association of Black Women Historians, and is a W. E. B. Du Bois Research Fellow with the University of Massachusetts at Amherst Du Bois Center. Dr. Donaldson earned her PhD in African American History from the University of Memphis and her BA and MA from the University of Tennessee at Knoxville.

LOUISA KOEBRICH taught courses in US history and the era of the American Civil War at West Point before assuming duties as a strategist at US Army North, where she develops future plans and programs in defense of the homeland. She earned her MA from Georgetown University and specializes in the history of African American service from the American Revolution through Reconstruction and the relationships between citizenship and military service. She is a coauthor of "'I Hope to Have Justice Done Me or I Can't Get Along Here': James Webster Smith and West Point," published in the *Journal of Military History*.

RONALD G. MACHOIAN is an Associate Dean in the International Division at the University of Wisconsin–Madison, where he also is a faculty associate who teaches global security courses in the International Studies major. He previously was faculty in the Military and Strategic Studies Department at the US Air Force Academy in Colorado Springs, where he also held leadership roles across cadet life and in international programs and education. He retired from the US Air Force in 2014 after having commanded at the squadron and group levels. He holds MA and PhD degrees in history from the University of Missouri–Kansas City. His scholarship focuses on the army's transition to a professionalized force during the nineteenth and early twentieth centuries. He is the author of *William Harding Carter and the American Army: A Soldier's Story.*

CAMERON D. MCCOY is an Asness Family Foundation Fellow in the Jenny Craig Institute for the Study of War and Democracy and a resident senior officer at the US Naval War College, Center for Naval Warfare Studies for the 2023–24 academic year. He earned a PhD in US history at the University of Texas at Austin after receiving a master's in military history at Texas A&M University and his bachelor's in International and Area Studies at Brigham Young University. He is the author of *Contested Valor: African American Marines in the Age of Power, Protest, and Tokenism.*

RORY MCGOVERN is a career US Army officer now serving as an Associate Professor of History at the US Military Academy, where he directs the American and military history programs. He earned a PhD from the University of North Carolina at Chapel Hill. He researches and writes about American military history from the Civil War through World War I. Among his past works are *George W. Goethals and the Army: Change and Continuity in the Gilded Age and Progressive Era;* with Makonen Campbell and Louisa Koebrich, he co-wrote, "'I Hope to Have Justice Done Me or I Can't Get Along Here': James Webster Smith and West Point," published in the *Journal of Military History.*

AMANDA M. NAGEL is an Associate Professor of Military History at the US Army Command and General Staff College's Department of Military History. She has also taught at the US Army School of Advanced Military Studies, the US Military Academy at West Point, and Winona State University in Minnesota. She earned her PhD from the University of Mississippi, specializing in African American history and global conflict. Her research centers on race, war, empire, masculinity, and citizenship in the United States at the turn of the twentieth century. She is currently revising a manuscript examining African American soldiers in the Spanish-American, Philippine-American, and First World Wars for the University of Virginia Press.

INDEX

The Black Soldier in War and Society

New Narratives and Critical Perspectives

This series is open to a wide array of scholarship on the ramifications of "soldiering" on the economic, social, cultural, or political lives of Black individuals, families, and communities. The editors seek projects that will highlight the long, and in many cases unending, fight for racial and social justice across time and space in the Black Atlantic.

9 780813 951911